EUROPEAN DISPOSAL OPERATIONS

The Sea Disposal of Chemical Weapons

H. LINDSEY ARISON III

Cover Design by: **www.ebooklaunch.com**

Maps at Figures 1, 54, 58-61, 77, 79, 82, 83, 93-96, 99, 103 and 104 courtesy of Google Earth Pro: **www.google.com/earth/index.html**

ISBN 13: 978-1490927657
ISBN: 1490927654

Preface

Epigraph

"All the declarations of independence in the world will not render anybody really independent. You may disregard your environment, you cannot escape it; and your disregard of it will bring you moral impoverishment and someday unpleasant surprises."

GEORGE SANTAYANA
Character and Opinion in the United States

Dedication

This work is dedicated to –

My Father
Major Harold L. Arison, Jr.
Who died prematurely in 1998 of causes attributable to
his exposure to the toxic defoliant agent orange while
serving in Vietnam in the late 1960s

and to

My Dear Wife
Wendie B. Arison
For her unfailing faith, inspiration, love, and devotion

Acknowledgements

I am very grateful to the many organizations and individuals who have provided me with crucial information for this study. Key agencies which have been most helpful include the:

- U.S. House of Representatives,
- U.S. Library of Congress,
- U.S. State Department's Bureau of Arms Control, Verification and Compliance, formerly the Arms Control and Disarmament Agency (ACDA),
- U.S. Environmental Protection Agency (EPA),
- National Archives of the United States,
- Pentagon Library, U.S. Department of Defense (DoD),
- Edgewood Arsenal Historical Office, Aberdeen Proving Ground, Maryland,
- Public Record Office, U.K.,
- Army Historical Branch, U.K. Ministry of Defence (MoD),
- Irish Sea Forum,
- NATO Committee on the Challenges of Modern Society (CCMS),
- Stockholm International Peace Research Institute (SIPRI),
- Norwegian Defence Research Establishment (NDRE), and the

- German Federal Armed Forces Scientific Institute for NBC Protection Technologies.

I am particularly grateful for the generous support provided by the staff of the former Department of Maritime Studies and International Transport, University of Wales, Cardiff, UK with very special thanks to Dr. Rhoda C. Ballinger, School of Earth and Ocean Sciences, Cardiff University.

Table of Contents

INTRODUCTION

"One thousandth of all mustard gas dumped – 27,000 kg – has a killing capacity of 5.4 million people".[1]

MAJOR GENERAL BORIS T. SURIKOV AND PROF. DR. EGBERT K. DUURSMA
Dumped Chemical Weapons in the Sea – Options

Prologue

The Chemical Weapons Convention (CWC), which opened for signature in January 1993 and entered into force on April 29, 1997, defines "Chemical Warfare Agent" and "Toxic Chemical" as: "any chemical which through its chemical action on life processes can cause death, temporary incapacitation, or permanent harm to humans or animals. This includes all such chemicals, regardless of their origin or of their method of production, and regardless of whether they are produced in facilities, in munitions, or elsewhere."[2]

"Chemical Weapons" means the following, together or separately:

(a) Toxic chemicals and their precursors, except where intended for purposes not prohibited under this Convention (CWC), as long as the types and quantities are consistent with such purposes;

(b) Munitions and devices, specifically designed to cause death or other harm through the toxic properties of those toxic chemicals specified in subparagraph (a), which would be released as a result of the employment of such munitions and devices;

(c) Any equipment specifically designed for use directly in connection with the employment of munitions and devices specified in subparagraph (b).[3]

"Old Chemical Weapons" (OCW), a subset of chemical weapons, are chemical weapons which were produced before 1925 or chemical weapons produced in the period between 1925 and 1946 that have deteriorated to such an extent that they can no longer be used as chemical weapons.[4] There are essentially four general categories of OCW:

1. Abandoned Chemical Weapons (ACW) – Chemical weapons, including old chemical weapons (OCW), abandoned by a State after 1 January 1925 on the territory of another State without the consent of the latter,
2. CW disposed of at sea or on land,
3. Discovered CW, and
4. OCW produced before 1946.[5]

Similar to the multifaceted problems posed by the disposal of nuclear waste, the cost-effective and environmentally safe destruction and disposal of chemical agents and munitions has posed a continuous problem since chemical weapons were first produced. Alternatives have included burning, land burial, or sea disposal. Regardless of the manner in which OCW were disposed, however, the problems related to them have virtually been ignored – and the potential environmental problems they now pose are significant.

Since the end of the First World War, dumping of chemical weapons at sea has taken place and is the subject of considerable concern in a number of international fora. Dumping of these materials has been reportedly carried out in every ocean, with considerable amounts in European bodies of water, principally the Baltic Sea.

Link to Video by the Monterey Institute on the Global Disposal of CW:
http://www.youtube.com/watch?v=wW76ha24QhA

Link to 1-Page PDF Summary: "Dumping Was Carried Out on a Global Scale":
"The U.S. Army unloaded its CW stockpiles off the coasts of more than a dozen countries as WWII ended"
http://www.dailypress.com/media/acrobat/2005-10/20226255.pdf

Global disposal operations will be addressed in detail in the second volume of this research: "Global Disposal Operations and Environmental Impact."

Marine dumped chemical munitions react differently in water depending on the agent they contain. The munition shell may break open during the dumping operation or may corrode over time, allowing the agent to leak out. Nerve agents and many other agents hydrolyze, or break down and dissolve once they come into contact with water, and are therefore rendered harmless in a relatively short amount of time. Mustard gas, however, is insoluble in water and most injuries that have occurred when fishermen come into contact with marine dumped chemical munitions have resulted from mustard gas.[6]

The full extent of this dumping will never be known due mainly to inadequate documentation of operations at the time of dumping and the subsequent loss or destruction of records that may have been taken. Remediation of marine chemical weapons dumpsites is technically challenging because of the nature of the material dumped and the uncertainty surrounding the quantities, type, locations, and the present condition or stability of these materials.[7] Exacerbating the problem is the fact that while little is known about the hydrolysis of chemical warfare agents concerned, nothing is known about their interactivity in seawater. Also, while there is a great deal of documentation on the acute effects of chemical warfare agents, very little is known about the delayed effects on human health and the environment.[8]

The history of modern chemical warfare began on April 22, 1915 when German forces attacked French and Canadian troops with chlorine gas at the Battle of Ypres. During World War I, a wide range of new agents was introduced. The discovery of a new class of chemical warfare agents (CWA) – the organophosphorous agents, including tabun (GA), sarin (GB), and soman (GD) – in the late 1930s and 1940s added a new dimension to chemical armament.[9]

Link to Video on the History of CW:
http://www.youtube.com/watch?v=5RUP3eiAuSw

Chemical weapons were only used in three known military operations during World War II. German forces are reported to have dropped mustard gas bombs on the suburbs of Warsaw, Poland on September 3, 1939. On May 7, 1942, German forces reportedly used chemical mortar bombs in Crimea and in the same month, used poison gas against people sheltering in underground tunnels in the stone quarries of Adzhimushkai near the ancient town of Kerch on the Crimean peninsula. Approximately 10,000 Soviet servicemen and civilians were poisoned and over 3,000 perished in the latter incident alone.[10] The 'Museum of the Heroic Defense of the Adzhimushkai Quarry in 1942', was established and dedicated to those who perished in the attack.

Despite the fact there was no large-scale use of CW during the Second World War, Germany, Japan, the Soviet Union, the United Kingdom, and the United States stockpiled massive amounts of these weapons in an arms race that was fueled by the perception that other countries possessed chemical weapons and intended to use them.[11] At the end of World War II, Germany had a large CW stockpile consisting mainly of mustard gas (H), nitrogen mustard (HN), soman (GD), and tabun (GA) and the World War I agents phosgene (CG), chloroacetophenone (CN), adamsite (DM), hydrogen cyanide (AC), and clark (DA/DC). Additional detail is provided in Chapter 2, Accounting of All Captured Chemical Weapons.

In 1945, at the conclusion of the war, the United States, United Kingdom, and Soviet Union set up the Allied Control Commission (ACC) to deal with the problems associated with the discovery, dismantling, and disposal of captured German CW stocks. In addition, there was a need to dispose of the chemical weapons and related equipment that had belonged to the U.K. and the former German allies, such as Hungary and Italy. Dumping at sea was the method of disposal which was chosen, and the environmental consequences of these activities were essentially ignored as the effect on the environment was not considered important at the time. The complexity of the issue was not given adequate consideration and the knowledge obtained from earlier disposal activities in Europe and elsewhere was ignored.[12]

The essence of the potential problem created by the ignorance and carelessness of that era is this: The majority of the CW munitions dumped in European waters were in artillery shells with shell casings that were several centimeters thick. Some agents, such as the nerve gas tabun, will have dissolved in the seawater if they have leaked out of their original casings, but most of the agents are probably still inside the shells or their original casings.[13]

What amount of toxic chemical agents remain? In the National Report of the Russian Federation entitled "Complex Analysis of the Hazard Related to the Captured German Chemical Weapons Dumped in the Baltic Sea" published in 1993, it is estimated that chemical agents account for 17.1% of the gross weight of the munitions.[14] The January 1994 "Report on Chemical Munitions Dumped in the Baltic Sea" prepared by the Ad-Hoc Working Group on Dumped Chemical Munition (HELCOM CHEMU) estimates 10-20%[15] In the study "The Challenge of Old Chemical Munitions and Toxic Armament Wastes" published by the Stockholm International Peace Research Institute (SIPRI) in 1997, it is estimated that chemical agents account for 10-25% of the gross weight of the munitions.[16]

Given these estimates and ranges (10-25%), if

approximately 754,975 tons (gross weight) of chemical weapons were sea disposed in European waters after WWII, it can therefore be estimated that between 75,497 and 188,744 tons of chemical agents (net weight) were disposed in this geographic area.

Estimates of total chemical agents sea-disposed in European waters post-WWII are addressed in Chapter 5.

The environmental and technical problems which have occurred since the end of World War II have demonstrated the risks that old chemical munitions pose to people and the environment. Many countries must cope constantly with newly discovered chemical weapons and conventional munitions from World War I (Belgium, France, Germany, and the U.K. are particularly affected). In many European countries, old CW munitions in the soil and water, and the contamination created by them, have been described as 'chemical time bombs'.[17]

The changed political climate in Europe at the end of the 1980s and the current interest of most European states and the European Union in environmental policy has led to the realization that there is a need for greater cooperation on environmental matters. Nevertheless, the destruction of chemical weapons does not appear to be a top priority for most European states. Moreover, because most of the information concerning the sea disposal of chemical munitions was classified, a knowledge base is virtually nonexistent. Access to reliable data on the number and location of sea disposal sites, and the quantities and types of munitions disposed have heretofore been restricted. This led to the publication of a number of speculative and sensational journal articles in the early 1990s such as "Dumps of Death"[18] and

"Scandinavia's Underwater Time Bomb".[19]

Inception and Scope

In May 1993, while serving as Military Legislative Assistant on the personal staff of then U.S. Congressman Dave McCurdy, U.S. House of Representatives, I fortuitously had the responsibility of reviewing the April 1993 "Interim Survey and Analysis Report on the Non-Stockpile Chemical Materiel Program" published by the U.S. Army's Program Manager for Non-Stockpile Chemical Materiel, U.S. Army Chemical Materiel Destruction Agency. In paragraph 2.5 of that report, Sea Disposal Survey, it was disclosed:

> "The Army has been informed of a survey being conducted for the U.S. Arms Control and Disarmament Agency (ACDA)[20] concerning chemical weapons that were disposed of by sea burial. Coordination has been made with ACDA to receive the results of its survey. Besides the sea disposal programs carried out by the United States and other nations in the late 1960s, Allied Forces were responsible for the disposal of 250,000 tons of German chemical munitions (total weight) by sea burial in 1945."[21]

Both intrigued and alarmed about this disclosure, a copy of the 19 January 1993 Arms Control and Disarmament Agency (ACDA) report entitled "Special Study on the Sea Disposal of Chemical Munitions" (hereafter referred to as the ACDA Study) was subsequently obtained through the ACDA Office of Congressional Affairs. Whereas the report was revealing, it was also disappointing in its accuracy and scope. To

8

date, a thorough study has not yet been made of the effect these thousands of tons of chemical agents will have on the environment and public health when they eventually enter the water and the sediment of the water bodies in question.[22]

Principally because most of the historical documentation pertaining to the sea disposal of chemical weapons during and after World War II had been classified by the governments of the principal dumpers (the U.S., U.K., and U.S.S.R.) as "FOR OFFICIAL USE ONLY", "CONFIDENTIAL", "SECRET", or "TOP SECRET", the subject has seldom, if ever, been addressed in academic texts. Minor footnote references to sea-disposed chemical munitions did appear in the early 1970s in the SIPRI study "The Problem of Chemical and Biological Warfare" – but the authors relied heavily on literature sources. It was not until declassified official documents started to become accessible in the Public Record Office (PRO), U.K., and reports such as the ACDA Study were published in the early 1990s that public awareness was heightened.

Given the unique marine pollution problems posed by the sea disposal of chemical weapons, the challenge presented at the outset of this study was to refine the study's scope. It was reasoned that by selecting a specific timeframe and geographic area, sufficient academic and scientific rigor could be applied to enable the postulation of an achievable solution benefiting all States concerned.

The timeframe selected begins on September 12, 1944 with The London Protocol – a TOP SECRET agreement reached among the U.S, U.K. and U.S.S.R. regarding the exercise of total control over Germany. Germany was to be divided into three occupation zones and Berlin into three sectors. Overall control would be exercised by an Allied Control Commission (ACC) composed of three Commanders-in-Chief.

It was initially envisioned the timeframe would end on December 8, 1949 with the 32nd and final meeting of the Control Commission for Germany (British Element) [CCG(BE)][23] Zonal Demilitarization Committee. This was the final meeting of the last remaining committee established to handle, inter alia, the disposition of captured German chemical weapons. **Research has shown, however, that large stocks of CW considered "excess" to national needs were likely sea disposed as late as 1990,** a subject which will be addressed in detail in Chapter 3, Accounting of All Sea-Disposed Chemical Weapons. For this reason, the ending timeframe extends until the last known CW stockpiles were sea disposed.

Note that in accordance with the Chemical Weapons Convention (CWC), any State that disposes of chemical weapons at sea after January 1,1985 must officially declare such dumping to the Organization for the Prohibition of Chemical Weapons (OPCW). Legal responsibilities of States will be discussed in Chapter 6.

Overview of the CWC:
http://www.opcw.org/chemical-weapons-convention/about-the-convention/

The geographic area selected includes all major European bodies of water in which CW dumping is known to have occurred during the selected timeframe:

- The Baltic Sea,
- The Skagerrak and Kattegat,
- The North Sea,
- The Norwegian Sea,
- The Atlantic Ocean,
- The Bay of Biscay, and
- The Mediterranean Sea.

Figure 1
Map of Europe
The Selected Geographic Area for the Study

Records also indicate that over one million tons of conventional munitions – plus some reported excess CW stocks – were also dumped in Beaufort's Dyke in the Irish Sea presenting its own set of unique environmental concerns. These areas, and information on CW dumpsites in every ocean, are addressed in Chapter 7, in Recommendations for Future Research.

The Allied Forces – principally the United States, United Kingdom, and Russia – conducted the majority of the dumping. Germany and France conducted sea disposal operations to a much lesser extent.

The 21 nation States now most potentially at risk as a result of the chemical weapons sea-disposed during the selected timeframe include the following:

1. Republic of Albania
2. Kingdom of Belgium
3. Bosnia and Herzegovina
4. Republic of Croatia
5. Kingdom of Denmark
6. Republic of Estonia
7. Republic of Finland
8. French Republic
9. Federal Republic of Germany
10. Hellenic Republic (Greece)
11. Ireland
12. Italian Republic
13. Republic of Latvia
14. Republic of Lithuania
15. Montenegro
16. Kingdom of the Netherlands
17. Kingdom of Norway
18. Republic of Poland

19. Russian Federation (principally Kaliningrad Oblast[24], between Lithuania and Poland on the Baltic Sea)
20. Kingdom of Sweden
21. United Kingdom of Great Britain and Northern Ireland

Hypothesis, Aim, and Assumptions

Since many of the ~100 hulks laden with CW and scuttled in European waters appear to be intact, the potential problem of toxins leaking into the marine environment is inevitable unless the submerged hulks are completely sarcophaged. The significant amount of CW dumped piecemeal in the Baltic Sea, however, present another challenge. Steps can be taken in the interim, however, by all States involved to ensure warning and disaster preparedness.

The aim of this study is to fill the knowledge void on this highly sensitive subject that has been obfuscated for over half a century.

The fundamental premise of this study is that in the 1940s – when the subject sea disposals occurred – dumping of toxic CW into the ocean was the preferred disposal method and was not an act of malice, malevolence, or ill will. The effect on the environment was simply not considered important at that time. It is therefore not the intent of this book to affix blame or culpability.

Research Methodology

Organizing the complex problem created by the sea disposal of toxic chemical munitions in European waters post-WWII into a form which would enable systematic study was the central task. The methodology employed in this study includes the following three sequential protocols summarized below:

- What chemical weapons were sea disposed, and where?
- What is the nature of the problem caused by the sea disposal, expressed in terms of potential environmental and public health impacts?
- What can be done to solve the problem?

1. Corroboration of the total amount of chemical weapons (CW) disposed in the selected geographic area during the selected timeframe.

In addition to the massive amounts of chemical warfare agents produced by the Third Reich between 1935 and 1945, the total tonnage found on German territory by the Allies included a significant amount of CW from Russia, France, Italy, and Hungary. While some discovered munitions and agents were destroyed, buried, or transferred to Allies' stockpiles (ostensibly "for research purposes"; in actuality to augment Allies' stockpiles), the majority were sea-disposed. And whereas the United Kingdom and the United States disposed of the majority of their captured CW by loading ships hulls and subsequently scuttling them in the Atlantic Ocean, Bay of Biscay, North Sea, Norwegian Sea, and the Skagerrak and Kattegat Straits, Russian forces typically dumped their captured CW piecemeal and randomly in predesignated Baltic dumping areas.

The strategy employed for this protocol was to obtain as many unclassified and declassified official government documents as possible from the principal dumping States: United Kingdom, United States, and Russia regarding the disposition of captured and national stockpiles of CW in the European Theater of Operations. "The History of Captured Enemy Toxic Munitions in the American Zone, European Theater, May 1945 to June 1947" published in 1947 by the Office of the Chief of Chemical Corps, United States Army European Command, is an example of such a resourceful document. Data obtained from these principal sources were compared with research conducted by numerous international governmental and civil agencies and organizations (e.g., The Danish National Environmental Protection Agency's "Detailed Report on the Environmental, Health, and Security Issues Associated with the Dumping of Poison Gas Ammunition in Danish Waters" published in 1985), published literature, information from the media, and investigatory correspondence to compile the most reliable, consolidated data set.

2. Assessment of the potential problem caused by over five decades of CW sea disposal by consideration of potential environmental and public health impacts.

An assessment of any potential environmental impact must consider the behavior of individual warfare agents in seawater, their potential pathways, and threats posed. While comparatively little is known about the behavior of chemical agents in seawater and their hydrolysis products (which have, in some instances, equal or greater toxicity), virtually nothing is known about their interaction or synergies. A crucial caveat here is that data regarding hydrolytic detoxification of agents which are based on laboratory investigation may be inaccurate for agents that are located in bodies of water.

The research methodology employed for this element of the protocol was to obtain unclassified and declassified official government documents and scientific research conducted by international governmental and civil agencies and organizations regarding the behavior of individual warfare agents in seawater, their effects on the marine environment, and evidence of environmental impact. The German Federal Maritime and Hydrographic Agency's 1993 report "Chemical Munitions in the Southern and Western Baltic Sea" and the Stockholm International Peace Research Institute's 1985 monograph "The Detoxification and Natural Degradation of Chemical Warfare Agents" are examples of such documents. Information from the media has proven to be the primary source of evidence of environmental impact.

An assessment of any potential public health impact must consider the acute physiological, chronic, and delayed toxic effects of individual agents on humans and evidence of impact on public health to date. The research methodology employed for this element of the protocol was to obtain unclassified and declassified official government documents and scientific research conducted by international governmental and civil agencies and organizations regarding the effects of individual agents on humans. The Stockholm International Peace Research Institute's 1975 monograph "Delayed Toxic Effects of

Chemical Warfare Agents" is a key example of a primary document. Information from the media has proven to be the primary source of evidence of impact on public health.

Because of the magnitude of information now available regarding potential environmental impact – and since the impact of the sea disposal of chemical weapons is a critical global issue, this protocol will be addressed in detail in the second volume of this research: "Global Disposal Operations and Environmental Impact".

3. Development of a solution to address the problem.

A detailed analysis of principal findings underscores the imperative for an international strategy and a proposal for international collaboration and cooperation in addressing the potential problem is advanced.

EUROPEAN DISPOSAL OPERATIONS

CHAPTER 1

EVOLUTION OF PLANS FOR THE DISPOSITION OF CAPTURED CHEMICAL WEAPONS

When the sea disposal of significant quantities of chemical weapons in European waters began in 1945, there were no international instruments in force relating to the marine environment in general or pollution by dumping, in particular. The disposal of chemical weapons at sea was therefore an acceptable alternative; the effect on the environment was not considered important at the time.[25] It was not until the Convention on the High Seas was promulgated in 1958 that the general issue of sea dumping was dealt with in the international community.

In the 1970s, several additional agreements were concluded, reflecting the growing international awareness of damage done to the environment, including ocean pollution. In the United States, the Marine Protection, Research, and Sanctuaries Act of 1972 (MPRSA, "Ocean Dumping Act") (Public Law 92-532) outlawed U.S. dumping of chemical weapons. Nonetheless, dumping of chemical weapons was not explicitly addressed internationally until negotiations towards a Chemical Weapons Convention (CWC) occurred in the 1980s.

Understanding there were no legal restrictions or impediments at the time, the purpose of this chapter is to summarize how the Allies' plan for the disposition of captured chemical weapons evolved – from The London Protocol on September 12, 1944 until the first U.S. sea disposal convoy, which was conducted on July 1, 1946. A detailed chronology of events in the American and British zones can be found at Appendix C.

1.1 The Allies' Plans

Putting events into perspective, it is important to note that the D-Day invasion occurred on June 6, 1944. The first significant decision following the invasion relevant to the plan for the disposition of captured CW was The London Protocol of September 12, 1944 – an agreement (originally a TOP SECRET document)[26] reached among the Governments of the United States of America, the United Kingdom and the Union of Soviet Socialist Republics on the zones of occupation in Germany and the administration of "Greater Berlin".

Link to the declassified TOP SECRET document, The London Protocol:
http://docs.fdrlibrary.marist.edu/PSF/BOX32/T298E01.H TML

The agreement divided Germany into three sectors (U.S., U.K., and U.S.S.R.). While it was unanimous the Soviets would occupy the eastern region of Germany, the United States and Britain argued over which sector they would control. At stake was sea access.

Both nations wanted the north-western zone with its ports of Bremen and Bremerhaven to serve as logistical bases for their forces. The London Protocol also established an Allied Control Council (ACC) composed of three Commanders-in-Chief to exercise overall control.

Link to You Tube video: "Allies Sign Control Law 1945": **http://www.youtube.com/watch?v=ao6F0KUvlYU**

At the conference near Yalta, in the Ukraine, February 4-11, 1945, Prime Minister Winston S. Churchill, President Franklin D. Roosevelt, and Premier Joseph Stalin (shown in Figure 2 below) met to determine overall Allied strategy. Finalizing a key initiative of The London Protocol of September 12, 1944, the concept of zones of occupation was decided. [27] France was co-opted as the fourth controlling power and allocated its own occupation zone.

Figure 2
Prime Minister Winston Churchill, President Franklin D.
Roosevelt, and Premier Josef Stalin
Yalta Conference, February 4-11, 1945

In the final agreement reached at the conference, the Americans received the northern ports as well as access rights through the British sector to the harbors. This arrangement providing crucial rail access to northern ports was a significant development for the U.S. sea disposal program. Another important development was Stalin's concession to allow a French occupation sector. Initially opposed to the proposal, Stalin conceded on the condition that the new zone be created out of the American and British sectors. A map displaying the four zone apportionment is shown at Figure 3.

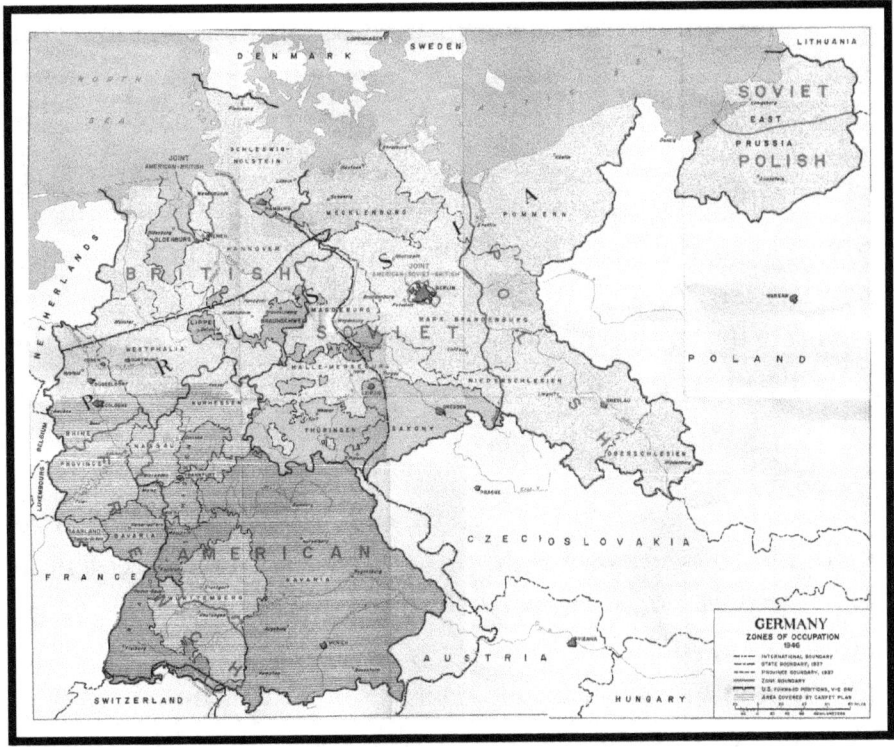

Figure 3
Zones of Occupation in Germany
Agreed at the Yalta Conference
February 4-11, 1945

The zones were governed by the aforementioned Allied Control Council, consisting of the four supreme commanders of the Allied Forces. The ACC's decisions were to be unanimous. If agreement could not be reached, the commanders would forego unified actions, and each would confine his attention to his own zone, where he had supreme authority. The ACC had no executive authority of its own, but rather had to rely on the cooperation of each military governor to implement its decisions in his occupation zone.

As a historical reference, on May 8, 1945, the unconditional surrender of Germany was formally ratified and confirmed in Berlin ending the war in Europe.

Despite the achievements of previous negotiations, by May, 1945 no clear agreement had yet been reached by the British, American, and Russian allies as to what was covered by the term "war material" in the context of the demilitarization of Germany. As a result, the British Element of the Control Commission for Germany [CCG(BE)], the agency responsible for administering the British Occupied Zone of Germany, noted that there was unlikely to be anything approaching a uniform treatment of German war material throughout all the Zones (American, British, Russian, and French).[28] In the absence of agreed quadripartite policy and the urgency of the situation, Zonal authorities were obligated to develop their own strategies.

As evidence of this development, the British Army of the Rhine's (BAOR) Disarmament Progress Report #3 of July 1945 noted that considerable thought had been given to the problem of finding the most economical and acceptable method of disposal. Of the three methods considered (dumping at sea in deep water, interring in deep and flooded mines, or industrial break-down), sea disposal was considered necessary for the bulk of the munitions in question. "After consideration of the possible methods of disposal, it was decided that disposal by deep sea dumping would prove the most practicable method for all ammunition which could be transported in safety. The most important factors influencing this decision were:

1. The availability of hulks of medium tonnage which could be scuttled with their cargoes.
2. Adequate rail services from and within the depots, and availability of suitable facilities at ports for direct loading from rail to ship, and
3. The close proximity of these ports (Kiel and Emden) to deep water in the Skagerrak."[29]

BAOR Disarmament Progress Report #3 also noted that "it is believed that all major depots (in the British Zone) have been discovered, the contents of the small dumps had been moved to the larger dumps (all of which are guarded), and there is a total of approximately 111,000 tons of enemy CW munitions in the British Zone."[30]

As a historical reference, U.S. forces gained control of their zone's first chemical weapons depot on March 29, 1945. The last of five such depots found came under U.S. control on May 3, 1945.

Between 29 Mar 45 and 3 May 45, five German Army and Air Force CW depots were found in the U.S. Zone (Frankenberg, Wildflecken, Grafenwöhr, Schierling, and St. Georgen) containing a total of approximately 122,409 tons of chemical weapons. This represents 38.2% of the total amount of CW captured by the Allies. It is important to note that in addition to chemical weapons of German manufacture, captured munitions included quantities from Russia, France, Italy, Czechoslovakia, Hungary, Belgium and The Netherlands. From these five depots, U.S. demilitarization operations were undertaken employing chemical decontamination, detonation, burning, and scuttling at sea.

Researching the evolution of U.S. plans, over 50 now declassified TOP SECRET, SECRET, and CONFIDENTIAL documents and messages transmitted over the period 2 May 1945 - 23 July 1946 were obtained. Images of the most significant of these in terms of information content follow – including individual assessments of significance to U.S. policy. All documents and messages are listed in Appendix C, "Detailed Chronology of Events in the American and British Zones and U.S., U.K., Russian, and German CW Sea Disposal Operations."

<u>2 May 1945</u>

Declassified SECRET Memorandum of a Telephone Conversation between Major General William Porter[31] (Chief, Chemical Warfare Service) and Brigadier General Alden H. Waitt[32] (Executive Officer, Chemical Warfare Service)

EXCERPT OF TELEPHONE CONVERSATION BETWEEN MAJOR GENERAL PORTER
AND GENERAL ALDEN H. WAITT ON 2 MAY 1945

*   *   *   *

Waitt: Fine. Okay. That's good. Now, another subject. Question as to the disposal of our own stock destruction chemical munitions: So far as Rowan is concerned, that would mean that bulk mustard and 4.2 shells loaded toxic chemicals. If the British don't want the bulk mustard, we'd like an idea as to what should be done with it.

Porter: We want the bulk mustard. And the toxic shell. And all bombs.

Waitt: How about the other 11.9 mustard in ton container?

Porter: Yes.

Waitt: You want it?

Porter: Yes.

Waitt: What are you going to do with all that mustard?

Porter: We hope there is a place for it in the Pacific.

Waitt: All right. Then the answer is that the bulk chemical and the 4.2 shells and all bombs should be returned, or sent direct to the Pacific, depending upon directives to be received later.

Porter: Yes.

Waitt: Now. Have you anything in as to whether any toxic chemical ammunition, bombs and shells should be left in the theater for use of occupational forces?

Porter: I see no reason for it.

Waitt: You see no reason. I check with you on that. Then you plan to receive stocks from theater?

Porter: Yes.

Waitt: My next question was would like an expression of opinion from you or General Montgomery as to disposition of enemy toxic chemical aircraft bombs and enemy shells. These aircraft bombs, enemy, and large quantities are now being turned up. We're getting lots of enemy toxic ammunition here. The enemy bombs with some modification of bombs by addition of spans and suspension locks, and by addition of necessary electric equipment in the airplanes can be used, or a new fuse can be designed.

Figure 4a
Declassified SECRET 2 May 1945 Telephone Conversation
between MG Porter and BG Waitt

Now, if such can be done and it appears to be practicable, should the enemy bombs and shells be shipped back unused, or should they be destroyed?

Porter: St. John is in charge of that problem, Alden. Have you talked to him?

Waitt: I haven't talked to St. John, no.

Porter: I think he has a scheme for disposal of such equipment. That's his job now.

Waitt: I see. I think General Montgomery might have an opinion as to whether or not he wants us to keep those bombs in case modifications can be made.

Porter: I will ask him.

Waitt: All right. If he notifies Rowan about that, I'd appreciate it.

Porter: Okay.

*   *   *   *   *

2

Figures 4b
Declassified SECRET 2 May 1945 Telephone Conversation
between MG Porter and BG Waitt

Significance: The U.S. had stocks of chemical weapons in the European Theater. In this telephone conversation, BG Waitt was asking MG Porter what should be done now with the stocks of U.S. chemical weapons in the European Theater and what specifically should be done with the bulk mustard, if the British don't want it. The U.S. policy, as stated by MG Porter was – we want all of it shipped back. When BG Waitt asked "what are you going to do with all that mustard", the response was "We hope there is a place for it in the Pacific (Theater)". The resolution was: all bulk chemicals, bombs, and 4.2" chemical mortar shells were to be returned to the U.S. or sent directly to the Pacific, depending on directives to be received later. BG Waitt then asked if any U.S. CW should be left in Theater for use by the occupational forces. The resolution was: no U.S. CW will be left behind. BG Waitt then asked what should be done with captured "enemy toxic chemical aircraft bombs and enemy shells". MG Porter stated that Colonel Adrian St. John [Chemical Advisor, G-3, Supreme Headquarters, Allied Expeditionary Forces (SHAEF)] was in charge of that problem and that he had a scheme for the disposal of such equipment.

22 days later, in a 24 May 1945 declassified SECRET message from the Chemical Warfare Service to HQ, Communications Zone, European Theater of Operations, Paris, France – and referencing the 2 May conversation between MG Porter and BG Waitt – the following disposition of captured German CW was directed: The U.S. requests highest priority shipment of 3,000 German KC 250 III Tabun-filled (GA) aerial bombs and 5,000 German 105mm Tabun-filled (GA) artillery shells to

Edgewood Arsenal, Maryland. The message also requests the immediate air shipment of 5 KC 250 III Tabun-filled (GA) aerial bombs for the purpose of filling U.S. 4.2" chemical mortar shells with Tabun (GA).

24 May 1945

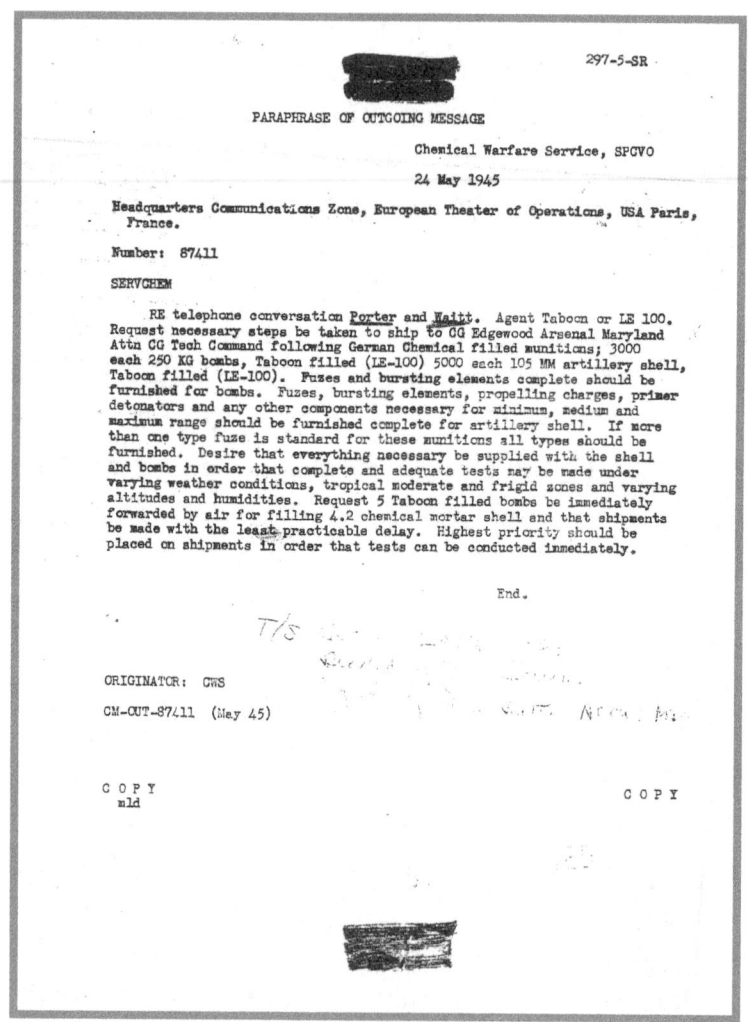

Figure 5
Declassified SECRET 24 May 1945 Message
Immediate Shipment of Tabun-Filled German CW to U.S.

9 July 1945

On 9 July 1945, the following declassified SECRET Message was sent from the CWS to HQ, Communications Zone, European Theater of Operations. Its significance is this: Because of an "urgent (U.S.) need for intelligence, technical examination, and field tests", the most expeditious shipment of 300 KC 250 German chemical-filled aerial bombs to Edgewood Arsenal, Maryland was directed. (100 filled with mustard (H), 100 with Phosgene (CG), and 100 additional filled with "LE-100") (LE-100 is a code name for Tabun.)

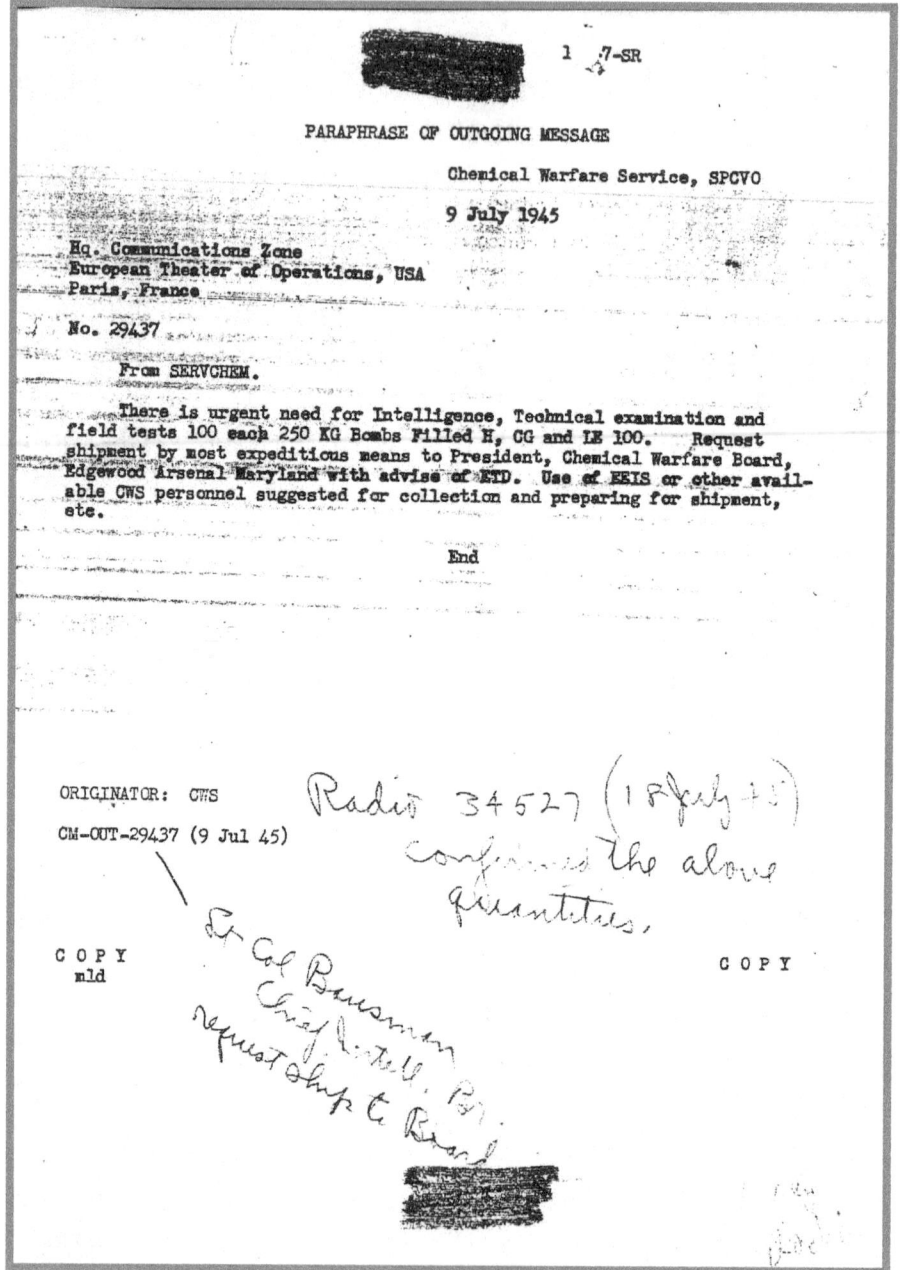

1 7-SR

PARAPHRASE OF OUTGOING MESSAGE

Chemical Warfare Service, SPCVO

9 July 1945

Hq. Communications Zone
European Theater of Operations, USA
Paris, France

No. 29437

From SERVCHEM.

There is urgent need for Intelligence, Technical examination and field tests 100 each 250 KG Bombs Filled H, CG and IE 100. Request shipment by most expeditious means to President, Chemical Warfare Board, Edgewood Arsenal Maryland with advise of ETD. Use of EEIS or other available CWS personnel suggested for collection and preparing for shipment, etc.

End

ORIGINATOR: CWS

CM-OUT-29437 (9 Jul 45)

Radio 34527 (18 July 45)
confirmed the above
quantities.

COPY
mld

Lt Col Bausman
Chief Intell B.
request ship to Read

COPY

Figure 6
Declassified SECRET 9 July 1945 Message
Immediate Shipment of 300 German KC 250 Aerial Bombs

With the consolidation of CW munitions at the U.K. depots complete in July 1945 and plans being made for sea disposal operations, it became clear to the War Office and the Admiralty that an organization needed to be formed on the continent to efficiently administer and manage the disposal programme.[33] Consequently, the Continental Ammunition Dumping Committee (CADC) was established and held its first meeting on July 15, 1945 at HQ, 21st Army Group.

The CADC was established by the Service authorities in London to centrally control the task of ammunition disposal and in particular to organize the dumping at sea of captured enemy ammunition not required for research or to meet the requirements of the London Munitions Assignment Board (LMAB).[34] (The LMAB was established to allocate specific items of German munitions, usually for research purposes, among the various members of the United Nations).

Although there was close liaison with the American forces, and that information was passed to the Norwegian, Danish, and Dutch authorities, the CADC was not an inter-Allied body. It was a British organization established solely to deal with ammunition in the British Zone; stocks under British control in Belgium, the Netherlands, and Denmark; and certain stocks of surplus British ammunition which it was desired to dump.

Note that the CADC was sub-divided into the CADC (Germany), which assumed responsibility for operations on the continent (the British Zone, Denmark, Belgium, and The Netherlands) and the CADC (UK) which handled the disposal of surplus British CW. Both the CADC

(Germany) and the CADC (UK) reported to the London-based Continental Movements and Shipping Committee (CMSC). According to the MoD, there is no evidence of any discussions of the CADC's program with the Russian authorities.[35]

At this first meeting of the CADC (15 Jul 45), it was agreed that the Army, through the 21st Army Group, would prepare the ammunition dumping program, including the program for CW disposal. Note that this program only concerned dumping where the use of scuttled ships was concerned; the CADC was not responsible for local arrangements for the use of small craft such as lighters for the dumping of loose conventional ammunition. It was also decided at this meeting that "suitable" sites for the disposal of mustard gas (Chemical Agent Code: H) munitions included an area off Stavanger [36] Norway at a depth of 100 fathoms (only 183 meters) and in the Kattegat.[37] No decision, however, could be made without further clearance from London.[38]

18 July 1945

The following declassified SECRET Message, transmitted on 18 July 1945, confirms the 300 German KC 250 aerial bombs the CWS requested on 9 July and clarifies the type requested: 100 KC 250 II bombs filled with thickened mustard/winterlost, 100 KC 250 II bombs filled with Phosgene (CG), and 100 additional KC 250 III bombs filled with Tabun (GA) (LE-100). These 100 Tabun-filled bombs are in addition to the 3,000 requested on 24 May.

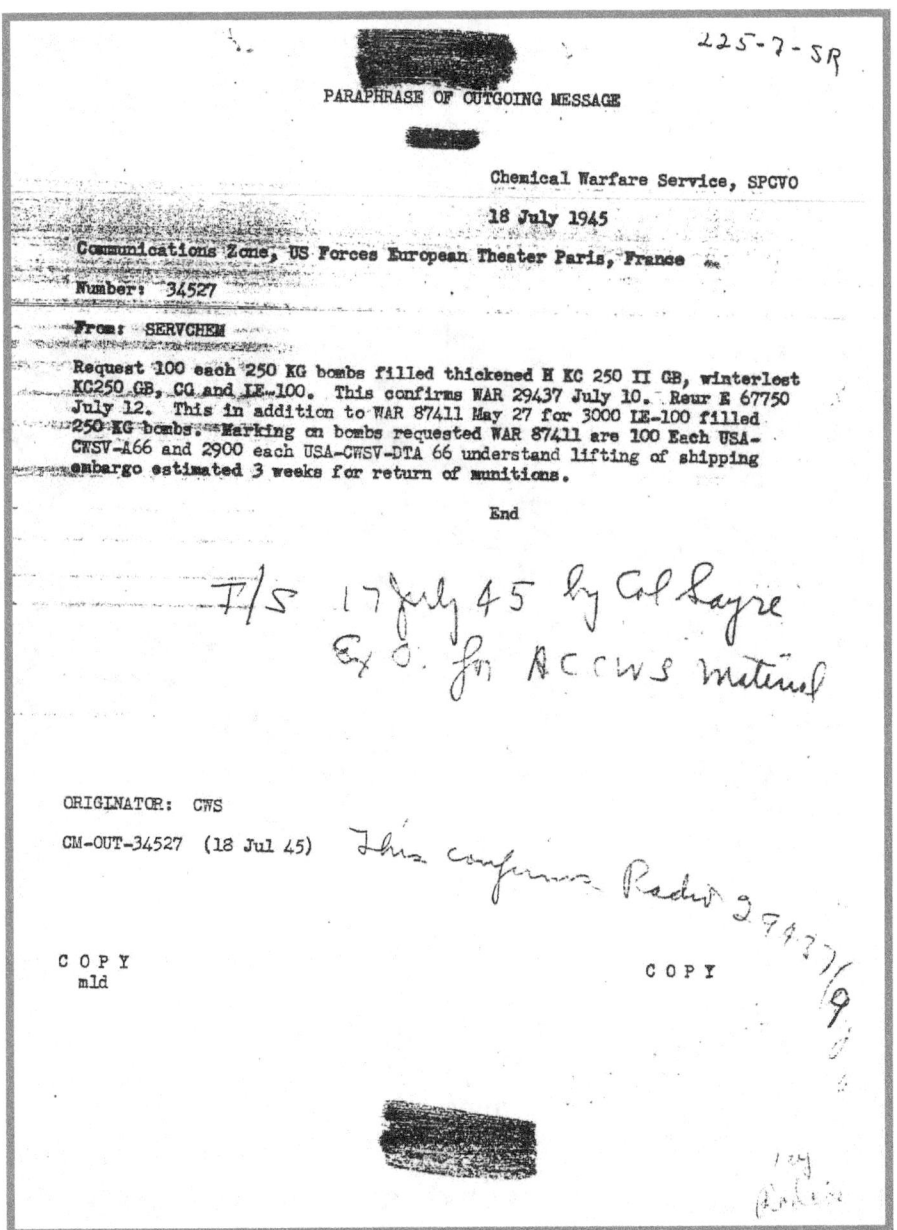

225-7-SR

PARAPHRASE OF OUTGOING MESSAGE

Chemical Warfare Service, SPCVO

18 July 1945

Communications Zone, US Forces European Theater Paris, France

Number: 34527

From: SERVCHEM

Request 100 each 250 KG bombs filled thickened H KC 250 II GB, winterlost KC250 GB, CG and LE-100. This confirms WAR 29437 July 10. Reur E 67750 July 12. This in addition to WAR 87411 May 27 for 3000 LE-100 filled 250 KG bombs. Marking on bombs requested WAR 87411 are 100 Each USA-CWSV-A66 and 2900 each USA-CWSV-DTA 66 understand lifting of shipping embargo estimated 3 weeks for return of munitions.

End

7/S 17 July 45 by Col Sayre
Ex O. for ACCWS material

ORIGINATOR: CWS

CM-OUT-34527 (18 Jul 45)   This confirms Radio 39437/

COPY
mld

COPY

Figure 7
Declassified SECRET 18 July 1945 Message
Requirement for 300 German KC 250 Aerial Bombs

At the Potsdam Conference, July 17 - August 2, 1945, the last of the World War II heads of state conferences, British Prime Minister Attlee, President Truman, and Premier Stalin (shown in Figures 4 and 5) met to clarify and implement agreements previously reached at the Yalta Conference.

Figure 8
Prime Minister Attlee, President Truman, and Premier Stalin
Potsdam Conference
July 17 – August 2, 1945

The outcome of the meeting, The Berlin Protocol of August 2, 1945 set out guidelines for the "complete disarmament and demilitarization of Germany" and further stated that "all arms, ammunition, and implements of war shall be held at the disposal of the Allies or destroyed."[39] This agreement formed the basis for the dumping of chemical weapons which took place after the war.[40] The detailed discussions of this subject were delegated to the Inter-Allied "Allied Control Commission" and its various subcommittees, the overall controlling body being the Allied Control Council (ACC) headquartered in Berlin.

Links to You Tube Videos Pertaining to the Potsdam Conference:
http://www.youtube.com/watch?v=2e4db-tHfZQ
http://www.youtube.com/watch?v=tYFjBmMIcbk
http://www.youtube.com/watch?v=8D3hw-f3Fp8

In the SECRET protocols of the Potsdam Conference, another inter-Allied body was created to facilitate the demilitarization process – the Tripartite Naval Commission (U.S., U.K., and U.S.S.R). Its task was to organize the disposal of the vessels of the German fleet. The Commission's decisions on these disposals required in many cases the scuttling of German naval vessels in deep water. Some could be used for the CW dumping program, but others which were not were in some instances scuttled within the areas used for CW-laden hulks, but not themselves carrying any CW munitions.[41]

Figure 9
The Potsdam Conference
July 17 - August 2, 1945

By August 1945, U.K. plans for sea disposal operations were now accelerating. At the third meeting of the CADC on August 16, 1945, the first details of the British Army's "C.W. Ammunition Dumping Programme" were developed.[42] The following table – which identifies depot numbers, locations, tonnage, loading ports, and shipping allocated – was included in the minutes of this meeting.

DRAFT

C.W. AMMUNITION DUMPING PROGRAMME

Serial	Depot No.	Location	Map Ref	Corps Dist	Tonnage	Loading Port	Shipping Allocated
1	50	MOLLN	S 9606	8	3900	SCHLUTUP	CLARA RUSS (2000) PILLAU
2	109	LEHRE	Y 0318	30	8000		(1800) TRITON (2000)
3	35	SCHNEVERDEN	S 3405	30	700		NEPTUNE (2000) LOUISE SCHRODER (1800) ALWINEROSS (1500) TRUDE SCHUNEMAN (1500) = 12,600
					12600		12600 @ 350 tpd = 36 days
4	24	MUNSTERLAGER	X 5890	30	16200	FLENSBURG	DRAU (8000) EDITH HOWALDT
5	29	CELLE	X599491	30	4500		(3000) BALKAN (3500) LOTTE
6	79	HANNIGSEN	X610350	30	600		(3500) DORA OLDENDORF (3500) = 21,500
					21300		21300 @ 450 tpd = 48 days
7	90	SENNELAGER	B 6853	1	3500	EMDEN	DUBORG (3500) KARL LEONHARDT
8	18	ESPELKAMPF	W 6020	1	11200		(8000) HELIOS (4000) OLGA
9	22	RAHDEN	W505478	30	3600		SIEMERS (5000) FALKENFELS
10	33	ORREL	X644880	30	13700		(10000) EMMY FRIEDRICH
11	78	BODENTEICH	X 9876	30	6500		(8000) = 38500
					38500		38500 @ 400 tpd = 97 days
12	38	WALSRODE	X253710	30	5500	NORDHAFEN	VICTORIA (8000) PATAGONIA
13	74	ILSTER	X 5393	30	8000		(8000) DER DEUTCHER (6000)
14	33	ORREL	X644880	30	19900		°JAN WELLENS (7000)
15	93	DIEKPUSEN	O 5092	30	3800		ˣSTETTIN (6000) JAN WELLENS
16	79	HANNIGSEN	X610350	30	1000		(3200) = 38200
					38200		38200 @ 400 tpd = 96 days
		Grand total			110,600		

°JAN WELLENS does one voyage @ 7000 tons to dump and scuttled on last voyage @ 3200 tons

ˣActs as accommodation ship to bring back crews of ships from scuttling area. On last voyage she loads to 6000 tons and is scuttled; other shipping provided to bring back crews.

Figure 10
Draft CW Dumping Programme from the
Minutes of the 3rd Meeting of the CADC, August 16, 1945

In the absence of policy guidance or direction from the Berlin-based Allied Control Council, the Combined Chiefs of Staff (American and British Chiefs of Staff) advised the respective commanders of the American and British Zones in Germany on August 17, 1945 that in the absence of any agreement by the Allied Control Council, both Zonal Commanders could investigate their own measures to destroy German war material in their respective areas.[43]

On 6 September 1945, Colonel Frank A. Bogart, Planning Division, Army Service Forces (ASF) chaired a meeting of the Ammunition Committee to review and discuss papers prepared as a result of the Committee's meeting on 30 Aug. Of key significance are the following:

1. A Memorandum for Record (MFR) prepared by Colonel Robert B. Judson, Ordnance Department, recommending unserviceable toxic munitions in both the European and Mediterranean Theaters of Operation by dumped at sea. This memorandum was transmitted on 7 September 1945 (See Figure 12).

2. A message prepared by Lieutenant Colonel C. L. Ogden, Distribution Division, listing specific items of toxic ammunition to be destroyed in theatre and items to be returned to the U.S. This message was also transmitted on 7 September 1945 (See Figure 13).

CHAPTER 1

6 September 1945

MINUTES OF AMMUNITION COMMITTEE MEETING

Held 1400, 6 September 45

1. Colonel Bogart opened the meeting by stating that the purpose was to review and discuss papers prepared as a result of the meeting of 30 August to be presented as follows:

    a. AGF - Lt. Col. Wing.

    b. Chief of Ordnance - Col. Judson.

    c. Distribution Division - Lt. Col. Ogden.

2. The AGF representatives submitted a name for the Chairman which contained the following information relative to the return of ammunition from overseas theaters:

    a. Types of ammunition which will not be required in the U.S. for training purposes.

    b. Types of weapons to be used in the U.S. for training purposes, and

    c. Ammunition items which should be given the lowest priority for return to the U.S., taking into consideration such factors as large stocks currently on hand, and the relatively small quantities required for the training program.

3. Col. Judson, Ord. Dept., submitted a letter prepared for dispatch to overseas theaters, which outlines the methods for disposal of surplus non-toxic ammunition. The methods contained therein were approved by all present, and referred to CWS representatives to add the disposal methods for surplus toxic ammunition, as Part II of subject letter, to be ready for dispatch by 10 September 1945.

4. Lt. Col. Ogden, Distribution Division, submitted a proposed radiogram for dispatch to oversea theaters, listing specific items of toxic ammunition to be destroyed in theater, and items to be returned to the U.S. This message was concurred in by those present and upon approval by G-4, WDGS, will be dispatched 7 September 1945.

5. Major Quicke, Requirements and Stock Control, presented two questions:

    a. Shall any toxic ammunition be returned to the U.S.

    b. Shall any unserviceable but repairable non-toxic ammunition be returned.

FILE

PRIORITY OF RETURN OF SUPPLIES

Figure 11a
Declassified CONFIDENTIAL
6 September 1945 Ammunition Committee Meeting Minutes

41

After discussion of the two questions, it was agreed that:

    (1) The following toxic ammunition will be destroyed locally; chemical bombs, handL filled 100 lb. M47 and 115 lb. M70, CK and AC filled 500 lb. M78 and 1000 lb. M79, all H, HS and L filled 4.2" chemical mortar shell.

    (2) All HT and HD ammunition including bulk will be returned.

    (3) Disposition covering CG, CN, DM and CNS cannot be determined pending completion of War Reserve requirements.

    (4) Unserviceable but repairable ammunition from 105mm to 240mm, will be returned, but 90mm and below, will be disposed of locally if in excess to theater requirements.

6. Results.

    a. Letter outlining disposal methods (3 above) is to be prepared by Chief, Chemical Warfare Service for dispatch by Director of Supply (Distribution Division) no later than 10 September 1945.

    b. Colonel Judson, Ordnance Department is to prepare a radiogram for dispatch by Distribution Division to all theaters, covering non-toxic items, unserviceable but repairable, excess to theater requirements, which are to be destroyed locally. In the preparation of this message, consideration is to be given to paragraphs 2, 4, and 5 above, and message is to be dispatched by 10 September 1945.

    c. Distribution Division will dispatch a radiogram to all theaters (4 above) on 7 September 1945, giving disposition of toxic ammunition, as outlined in 5 above.

2

Figure 11b
Declassified CONFIDENTIAL
6 September 1945 Ammunition Committee Meeting Minutes

d. The M6, M7, M12 and M13 Clusters and the M76 Incendiary Bomb will be considered for war reserves. This office has been advised by Capt. Duncan, your office, that the AAF will determine the order and priority of substitution.

e. Captured enemy munitions and materiel will be stored in present condition and modified when and if funds become available.

f. Bulk Napalm packaged in A-1 condition metal containers are considered suitable for long time storage when stored under suitable conditions and will be used for the war reserve. Igniters for fire bombs will be used for the war reserve.

g. Current ASF directives, Letter CG, ASF, to Chiefs of Technical Services, subject: "Special Instructions for Computation of MPR 20, 31 August 1945 file SPHLP 370.01, dated 27 August 1945, paragraph 6 l, prevents production to meet war reserves. Repair of material required to meet these reserves may be accomplished to the extent of fund availability. The priority for storage of 100 lb. incendiary bombs and 100 lb. persistent gas filled bombs for war reserve is:

(1) Complete rounds. Filled bombs with tail fins, fuzes and bursters.

(2) Filling for bombs.

(3) Empty bomb cases, complete balanced stocks. Cases, fins, fuze and bursters in proper proportion to make (1) above.

(4) Complete components except certain small parts which are dependent upon and have a short lead time to manufacture. An example is cases but not tail fins or bursters which can be made in a short time.

h. Present theater stockage of FS is to be disposed of in the theater in that it has been determined that there are ample stocks in the Z/I for known requirements. There are no known requirement for FM at this time.

i. It has been the policy of this Service to study each item when declared excess and to make recommendations for its disposition based on the requirements and the merits of the item and not to issue any general policy data.

FOR THE CHIEF, CHEMICAL WARFARE SERVICE:

JOHN C. MacARTHUR.
Colonel, CWS
Acting Director, Personnel, Plans & Training

cc: Industrial Div.
Technical Div.
Theaters Branch
Materiel Command
Supply Div, Material Plan.
Director, Personnel, Plans & Tng.

Figure 11c
Declassified CONFIDENTIAL
6 September 1945 Ammunition Committee Meeting Minutes

<u>7 Sep 45</u>

As approved at the 6 September 1945 meeting of the Ammunition Committee, on 7 September 1945 the Chemical Warfare Service transmitted the following memorandum directing the sea disposal of all unserviceable toxic munitions in the European and Mediterranean theaters of operation. The justification provided was this: "Due to hazards involved and excessive costs of handling toxics by any other method than dumping, recommend dumping at sea."[44] The CWS also recommended the immediate dispatch of a chemical officer with expertise in toxic munitions disposal to the theater to consult with the theater commander and to supervise disposal.

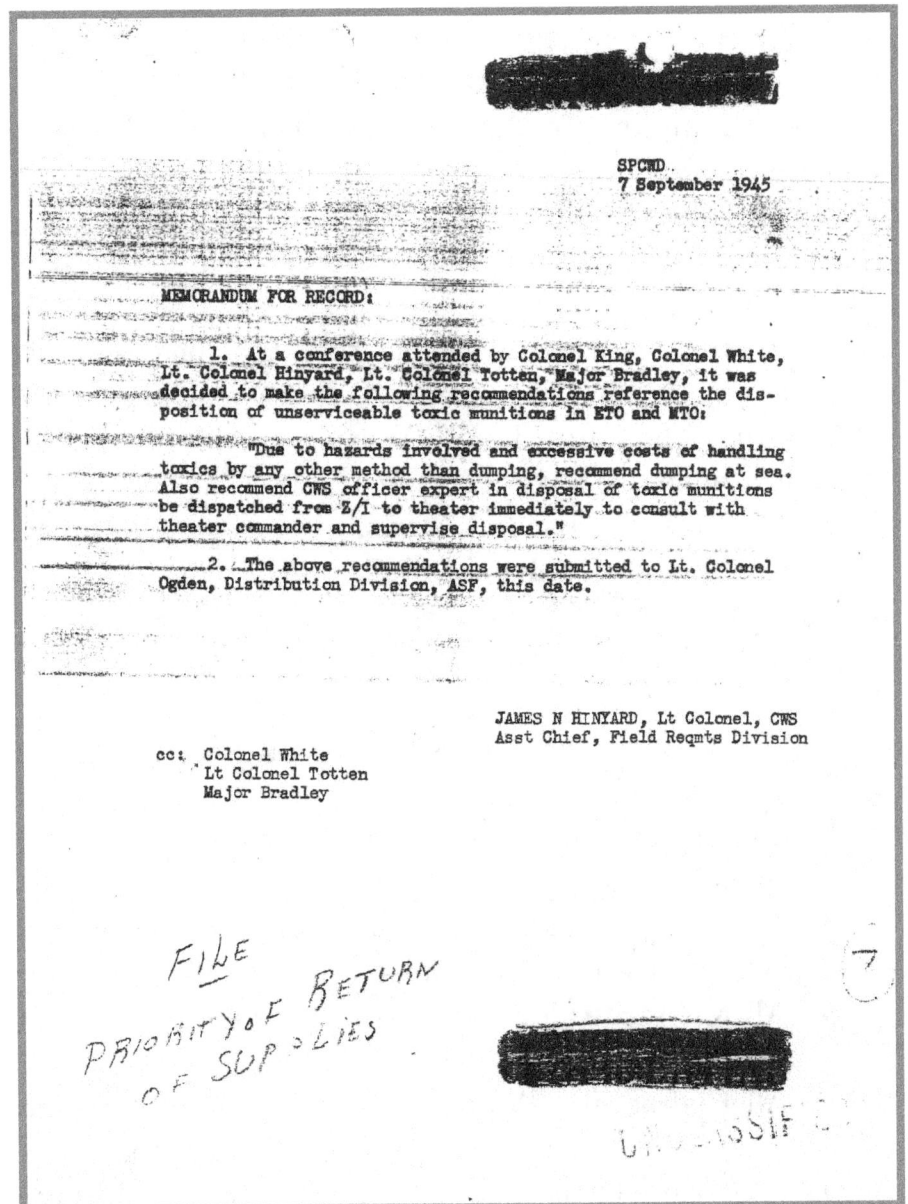

Figure 12
Declassified CONFIDENTIAL
7 September 1945 Memorandum
Directing Sea Disposal of Unserviceable Toxic Munitions

7 Sep 45

Also as approved at the 6 September 1945 meeting of the Ammunition Committee, the Chemical Warfare Service transmitted the following declassified CONFIDENTIAL message on 7 September 1945 detailing current policy for the disposition of captured toxic munitions.

Of immediate significance is that the message was transmitted to ALL Theaters of Operation: U.S. Forces, European Theater, Rear (Paris, France); USAF Mediterranean Theater of Operations (Caserta, Italy; Army Forces, Pacific Administration (Manila, Philippines); and U.S. Forces, India Burma Theater (New Delhi, India). The Commanding Generals of U.S. Ports of Embarkation at New York and San Francisco were also advised.

All Theater Commanders were directed to locally dispose of all of the following ammunition – except captured toxics that were in excess to theater requirements:

- All Mustard (H) and Lewisite (L) filled 100 pound M47 series chemical bombs
- All Mustard (H) and Lewisite (L) filled 115 pound M70 series chemical bombs
- All Hydrocyanic Acid (AC) and Cyanogen Chloride (CK) filled 500 pound M78 series chemical bombs
- All Hydrocyanic Acid (AC) and Cyanogen Chloride (CK) filled 1,000 pound M79 series chemical bombs
- All Mustard (H), Sulfur Mustard (HS), and Lewisite (L) filled 4.2" chemical mortar shells (except certain lots which were Thickened Mustard (HT) filled)

- All Mustard (H), Sulfur Mustard (HS), and Lewisite (L) Bulk Agents
- All toxic-filled artillery shells (Remember that 5,000 Tabun-filled (GA) artillery shells were requested on 24 May 45 to be shipped to Edgewood Arsenal, Maryland)

Sea disposal of CW was again cited as the preferred disposal method "since hazards involved and excessive handling costs limit other methods".

It was also noted that the 7 September 1945 instructions concerning the disposition of excess or surplus toxics supersede any previous disposition instructions.

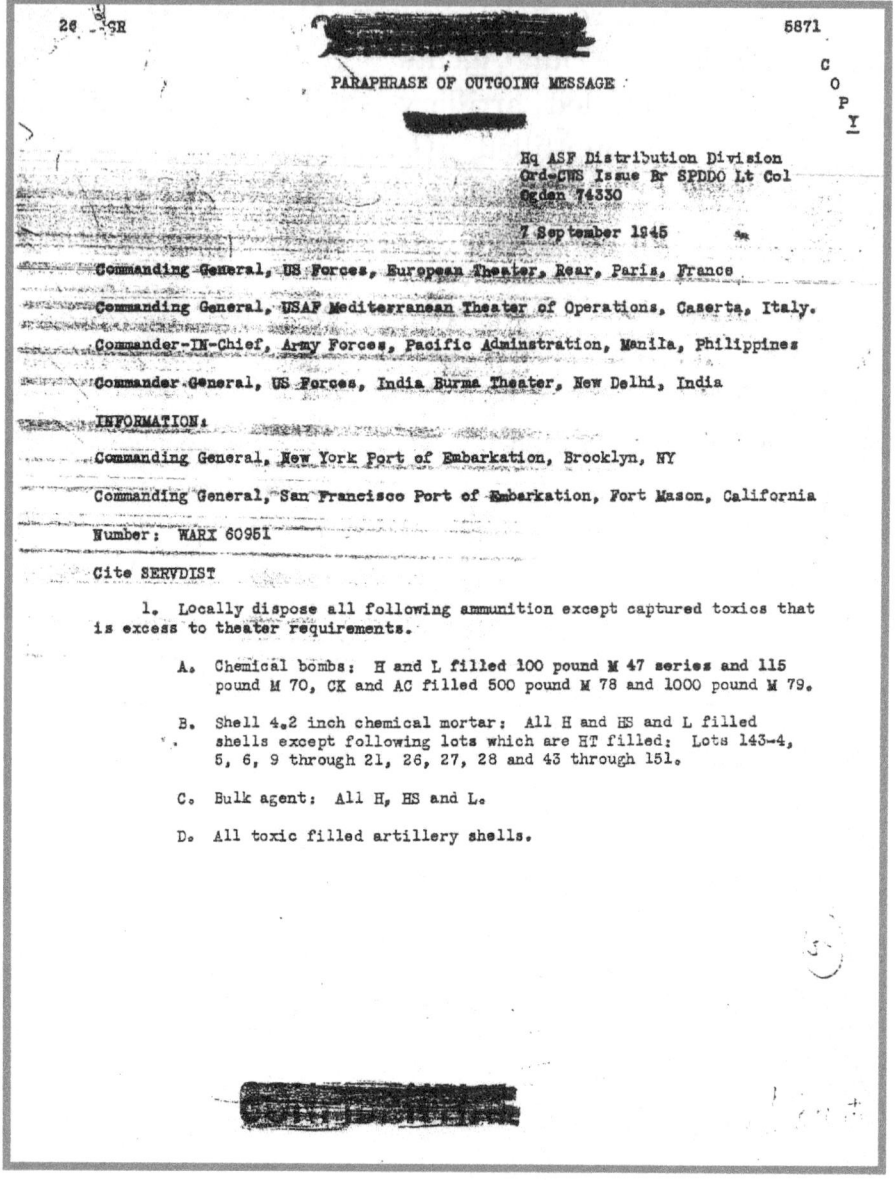

26 ~~GR~~   5871

PARAPHRASE OF OUTGOING MESSAGE

C
O
P
Y

Hq ASF Distribution Division
Ord-CWS Issue Br SPDDO Lt Col
Ogden 74330

7 September 1945

Commanding General, US Forces, European Theater, Rear, Paris, France

Commanding General, USAF Mediterranean Theater of Operations, Caserta, Italy.

Commander-IN-Chief, Army Forces, Pacific Administration, Manila, Philippines

Commander-General, US Forces, India Burma Theater, New Delhi, India

INFORMATION:

Commanding General, New York Port of Embarkation, Brooklyn, NY

Commanding General, San Francisco Port of Embarkation, Fort Mason, California

Number: WARX 60951

Cite SERVDIST

1. Locally dispose all following ammunition except captured toxics that is excess to theater requirements.

A. Chemical bombs: H and L filled 100 pound M 47 series and 115 pound M 70, CK and AC filled 500 pound M 78 and 1000 pound M 79.

B. Shell 4.2 inch chemical mortar: All H and HS and L filled shells except following lots which are HT filled: Lots 143-4, 5, 6, 9 through 21, 26, 27, 28 and 43 through 151.

C. Bulk agent: All H, HS and L.

D. All toxic filled artillery shells.

Figure 13a
Declassified CONFIDENTIAL 7 September 1945 Message
to All Theater Commanders
Disposition Instructions for Toxic Chemical Munitions

Book Message   Page 2   7 September 1945

    2. Above munitions, except lot numbers indicated for 4.2 inch chemical mortar are unsuitable for long term storage.

    3. 1 ton containers of H, HS and L will be emptied, retained and used to prepare all available excess bulk HT and HD for return.

    4. Report excess quantities grade 1, covered by lots in B above and bulk HT and HD in your theater so disposition instructions can be promptly furnished. Other than grade 1 4.2 inch chemical mortar ammunition will be disposed of locally.

    5. Prefer disposal by dumping at sea since hazards involved and excessive handling costs limit other methods. Complete instructions covering disposal methods will issue about 15 September.

    6. Instructions herein supersede any previous disposition instructions covering excess or surplus toxics.

    7. Separate disposition will follow, upon completion war reserve requirements, covering all CG, CN, DM and CNS ammunition, grenades and bombs. Policy covering captured toxics by CCS who will advise decision now under consideration.

               End

ORIGINATOR: ASF Dist

INFORMATION: ASF Plan-Supply-Ord-CWS-Trans-R&SCD-P10p
             CG AAF
             OPD
             G-4

CM-OUT-60951 (Sept 45) DTG 072029Z ls

Figure 13b
Declassified CONFIDENTIAL 7 September 1945 Message
to All Theater Commanders
Disposition Instructions for Toxic Chemical Munitions

<u>30 Sep 45</u>

Bringing the evolution of the British plan for the disposition of captured German CW to its conclusion is the September 30, 1945 report of the first British CW convoy contained in the 15th progress report of the Control Commission for Germany (British Element):

"The first convoy of five ships carrying approximately 17,000 tons of C.W. ammunition, which all belonged to the German Army, has sailed for the Skagerrak scuttling area. The loading of ships and hulks is now proceeding at the rate of about 2,000 tons a day."[45]

Army Division: "27,000 tons of CW ammunition has been dumped at sea out of a total so far uncovered of 120,000 tons."[46]

The MoD's "Report on Sea Dumping of Chemical Weapons by the United Kingdom in the Skagerrak Waters Post World War II" corroborates this data. The five ships in Convoy 1 (CW1) were the Duborg (3,500 tons of CW), Louise Schroeder (1,800 tons), Patagonia (8,000 tons), Pillau (1,800 tons), and the Triton (2,000 tons) – accounting for a total of 17,100 tons.

The estimate of 120,000 tons "uncovered" made in September 1945 was also accurate. The BAOR Disarmament Progress Report of December 1947, which was the last BAOR report to include CW dumping information, confirmed that 119,910 tons of CW munitions had been disposed of from British control.

<u>5 Oct 45</u>

In the evolution of U.S. plans, the declassified 5 October 1945 TOP SECRET telegram from the Joint Staff Mission in Washington to the Cabinet Offices was a milestone. It communicated the policy decision that quadripartite agreement on the disposal of German CW stocks is unnecessary, that the U.S. planned to fill any requirements for CW material from stocks available in the U.S. zone of occupation, and authorized the War Office to similarly fill any U.K. requirements from captured stocks in the British zone.

TOP SECRET

1. The United States Chiefs of Staff are informed that the Control Council (Germany) has agreed that the disposal of enemy war material will be the separate responsibility of each Zone Commander. In view of this fact, it is not considered useful to iniate action at this late date to secure quadripartite agreement to the policy on disposal of German chemical warfare stocks.

2. It is the intention of the United States Chiefs of Staff to fill as far as practicable U.S. requirements of chemical warfare material from stocks available in the U.S. Zones of occupation. There is no objection to making available to the United Kingdom under arrangements agreeable to the Commanding General, U.S. Forces, European Theatre any remaining stocks that may be desired for the British use, and it is suggested that such British requirements be made known directly to the Commanding General, U.S. Forces, European Theatre. Stocks of chemical warfare material remaining after the above requirements have been met will be destroyed.

Figure 14
Declassified TOP SECRET
5 October 1945 Telegram from the Joint Staff
at Washington to the Cabinet Offices

It is clear from the telegram above that:

- The disposal of captured war material was to be the separate responsibility of each Zone Commander,
- Agreement among the U.K., U.S., Russia, and France regarding policy on disposal of German chemical warfare stocks was hereafter considered unnecessary.
- The United States intended to transfer any "required" CW from captured stocks in the U.S. Zone to the U.S. inventory.
- The United Kingdom was welcome to claim any remaining CW in the U.S. Zone for the U.K. inventory, and
- Any CW remaining after U.S. and U.K. needs were met, would be destroyed.

18 Oct 45

As a logical follow-on to the Telegram from the Joint Staff Mission at Washington to the Cabinet Offices on 5 Oct 45, a conference was held on Thursday, 18 October 1945 for the purpose of making recommendations to Major General Porter (the Chief, Chemical Warfare Service) as to which types of captured German chemical bombs or bulk toxic agents should be placed in the War Reserve. Attendees included representatives from the Field Requirements Division, Technical Division, Supply and Distribution Division, Medical Division, and the Air Chemical Office. The declassified CONFIDENTIAL minutes of the Conference are at Figure 15.

War Reserve fill priorities were recommended for lung poison/choking agent Phosgene (CG) (Paragraph 4) and blistering agent Mustard (H) (paragraph 5). Considered "unsatisfactory", justification for NOT including the nerve agent Tabun (GA) in the War Reserve was provided (paragraph 3).

19 Oct 45

The declassified CONFIDENTIAL memorandum at Figure 16 signed by Brigadier General Alden H. Waitt, Assistant Chief, Chemical Warfare Service for Field Operations, summarizes the munitions and bulk stocks the U.S. considers as requirements for the War Reserve and directs their shipment back to the United States:

- All U.S. theater stocks of Phosgene-filled (CG) M78 and M79 Bombs
- All U.S. theater stocks of bulk agent Phosgene (CG)

- All U.S. theater stocks of bulk British Thickened Mustard (HT) and Distilled Mustard (HD)
- All stocks of the following German CW captured in the American Zone:
  - All Phosgene bombs (CG)
  - All bulk Phosgene (CG)
  - All Sulfur Mustard bombs (TGH) (Thiodiglycol Process)
  - All bulk Sulfur Mustard (TGH) (Thiodiglycol Process)

$7\ \text{uli}$

23 October 1945

MINUTES OF CONFERENCE ON WAR RESERVE OF CHEMICAL BOMBS AND TOXIC BULK AGENTS

1. A conference was held Thursday, 18 October 1945, in Room 2438 for the purpose of preparing recommendations to the Chief, CWS, as to which types of captured German chemical bombs or bulk toxic agents should be placed in the War Reserve. Representatives of Field Requirements Division, Technical Division, Supply and Distribution Division, Medical Division, and the Air Chemical Office were present as follows:

Colonel D. R. King, Field Requirements Division
Lt. Colonel C. W. Baber, Field Requirements Division
Lt. Colonel J. S. Entriken, Air Chemical Office
Lt. Colonel J. N. Hinyard, Field Requirements Division
Major R. J. DeGray, Technical Division
Major R. L. Fox, Technical Division
Major C. W. Knight, Field Requirements Division
Major J. J. Troy, Supply and Distribution Division
Major E. E. Turner, Supply and Distribution Division
Captain Duncan, Air Chemical Office
Captain J. O. Hutchins, Medical Division

2. GENERAL DISCUSSION

The purpose and definition of the War Reserve was explained in order that all representatives present would understand the ground rules under which this reserve is to be established. It was pointed out that no approval exists at present for production of any item to fill a War Reserve requirement. However, it is the policy now to retain all items which are available provided they are considered suitable to meet the War Reserve demand. Items with a comparatively short life are to be considered since during the period of their life they provide insurance against a possible need. For example, the M47 bomb may deteriorate within 6 months and it would be used for that period but used only if sufficient quantities of a more storable bomb were not available to completely meet the War Reserve requirement. However, it was agreed in the conference that where items had to be shipped back from overseas, the maximum storable life of the item should not be less than 3 years. With these general principles in mind, the various type agents and the priority of their use for War Reserve were considered, particularly where captured German munitions and agents were involved.

3. TABUN

a. Discussion. Technical Division representative presented a chart showing the storage stability of Tabun (copy attached). This chart

18 Oct 45 - Conference

Figure 15a
Declassified CONFIDENTIAL
18 October 1945 Meeting Minutes
Conference on War Reserve of Chemical Bombs
and Toxic Bulk Agents

was developed by applying theoretical equations to the limited data available. According to this chart, the storable life of Tabun at 65°C (149°F) would be 3 months; at 50°C (122°F), 7 months; and at 25°C (77°F), 43 months. At some of the storage areas, for example Deseret, it was estimated that the temperature might get as high as 120°F to 125°F for extended periods of time. It was estimated that if this material were stored in igloos, the temperature might not exceed 70°F. However, representative of Supply Division believed that required quantities of this material could be stored in igloos only if large quantities of some other material, such as HE chemical mortar shell, were stored outside. It was pointed out that the storage data on Tabun was based on tests on artillery shell. However, it was considered the same data would apply to bombs since the ratio of volume of liquid and metal surface contact would not be expected to influence the decomposition by polymerization. On the basis of this information, it was agreed that probably the maximum storage life of this material would be about 28 months under expected storage conditions and the date that these munitions were produced was not known.

Representatives from Medical Division explained the toxicological value of the agent, both before and after decomposition. It was stated that before decomposition this agent was not considered as more satisfactory than CG. Medical Division has no data on the effect of contamination of the skin. However, it was stated that in general for other agents, skin contamination has not been as effective as vapor. The polymerized agent would have practically no vapor pressure. To summarize, it was stated that in the opinion of those present the short life in storage of Tabun would not justify its return for the War Reserve. The possibility of the value of this agent as a persistent agent was discussed since shortages of both persistent and nonpersistent agents will probably exist. However, there is no indication that the residue after decomposition, that is, the polymerized agent, is of any value as a vesicant agent.

    b. Recommendations. That, based on the available data, Tabun not be considered satisfactory for the War Reserve.

    4. C.G.

        a. Discussion. The question of whether German CG and German CG bombs should be used for War Reserve was discussed. It was explained that Technical Division has not had the opportunity to examine samples of German CG. However, a study of the manufacturing procedures indicated that German CG is even more stable than U.S. CG. It was pointed out that although it would be desirable to have this data, the decision must be made without it, since the German material must be either shipped back or destroyed now while personnel is available in the theaters. It was also pointed out that the German bombs would have to be modified before they could be dropped from U. S. planes. It was stated that this modification was feasible and that the Air Forces had concurred in this modification immediately after VE Day. AAF representatives stated that this concurrence, as far as they knew, would apply to the present question. Funds may not be available for this modification; however, when the need for the War Reserve actually arises, it was

2

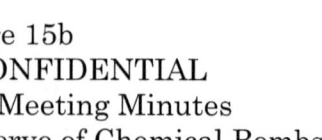

Figure 15b
Declassified CONFIDENTIAL
18 October 1945 Meeting Minutes
Conference on War Reserve of Chemical Bombs
and Toxic Bulk Agents

considered that a modification could be accomplished in a much shorter time than the manufacture and filling of bombs. It was stated that CG will store indefinitely provided it is free of water and since the German CG is considered to be free of water, the storable properties would be entirely satisfactory.

    b. Recommendations. That CG bulk agent and CG bombs be used to meet the demand for War Reserve in the following priorities:

    (1) Stocks in Z/I.

    (2) Stocks in ETO and MTO.

    (3) Captured German stocks.

    (4) Pacific stocks.

    (5) Z/I stocks of CK and AC bombs be used in the War Reserve only if a requirement exists for nonpersistent agents after utilizing all stocks of CG.

5. H

    a. Discussion. It was proposed that persistent gas requirements for War Reserve be filled in the following priority:

    (1) Levenstein H after distillation.

    (2) U. S. stocks of British HT.

    (3) Captured stocks of German TGH (pure H made by the thiodiglycol process).

    (4) M70 bombs in Z/I (filled with HD).

    (5) M47 bombs in Z/I (filled with HD).

It was pointed out that the German bombs filled with TGH are 250 and 500 kg bombs which are not considered as a desirable size. However, AAF representatives stated that if shortages exist, these bombs would be desirable. It was also agreed that these bombs would be placed in a higher priority than the M70 and M47 bombs in the Z/I, but in a priority below the bulk stocks of HD, HT, or TGH. There was considerable doubt as to whether any German bombs filled with TGH actually existed, at least, in any appreciable quantities. The demand should be stated, however, since there is no complete inventory of captured German material. It was agreed to set up the priorities even though no material may actually exist. The question was raised as to whether

3

Figure 15c
Declassified CONFIDENTIAL
18 October 1945 Meeting Minutes
Conference on War Reserve of Chemical Bombs
and Toxic Bulk Agents

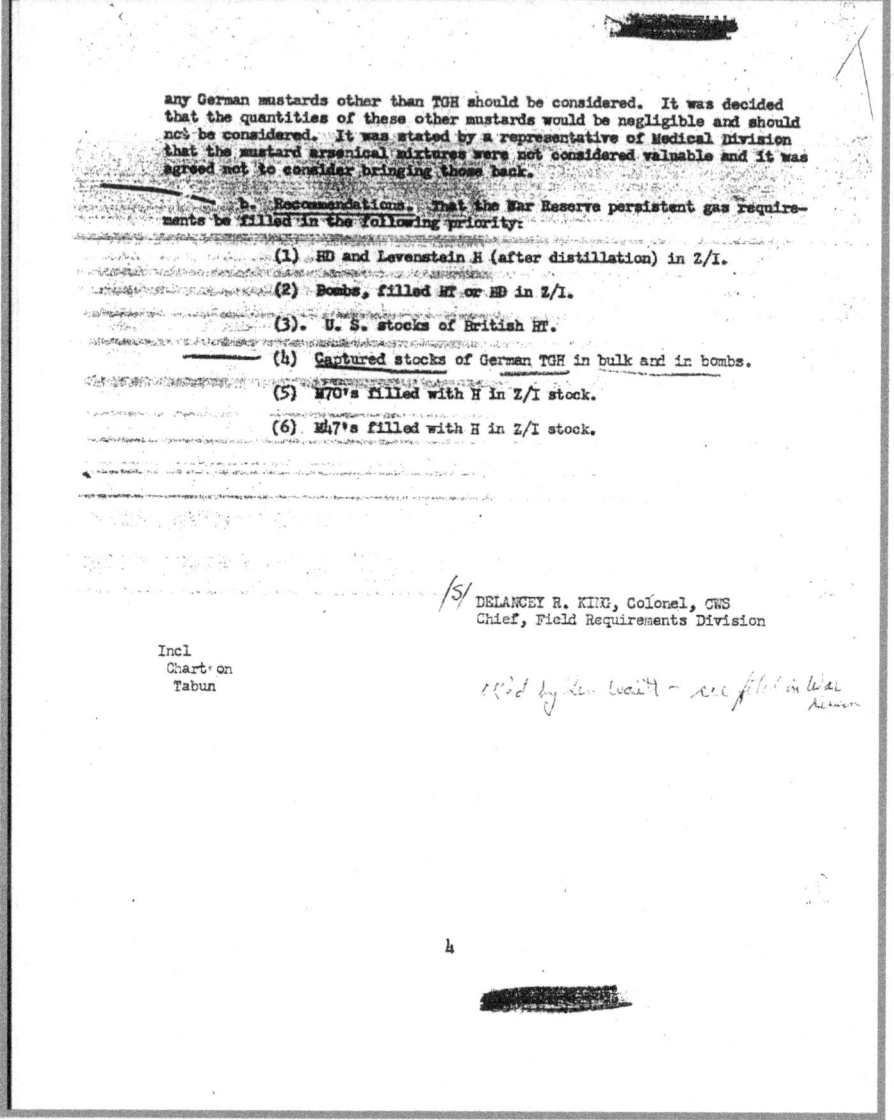

any German mustards other than TGH should be considered. It was decided that the quantities of these other mustards would be negligible and should not be considered. It was stated by a representative of Medical Division that the mustard arsenical mixtures were not considered valuable and it was agreed not to consider bringing them back.

b. Recommendations. That the War Reserve persistent gas requirements be filled in the following priority:

(1) HD and Levenstein H (after distillation) in Z/I.

(2) Bombs, filled HT or HD in Z/I.

(3). U. S. stocks of British HT.

(4) Captured stocks of German TGH in bulk and in bombs.

(5) M70's filled with H in Z/I stock.

(6) M47's filled with H in Z/I stock.

/S/ DELANCEY R. KING, Colonel, CWS
Chief, Field Requirements Division

Incl
Chart on
Tabun

4

Figure 15d
Declassified CONFIDENTIAL
18 October 1945 Meeting Minutes
Conference on War Reserve of Chemical Bombs
and Toxic Bulk Agents

19 Oct 45

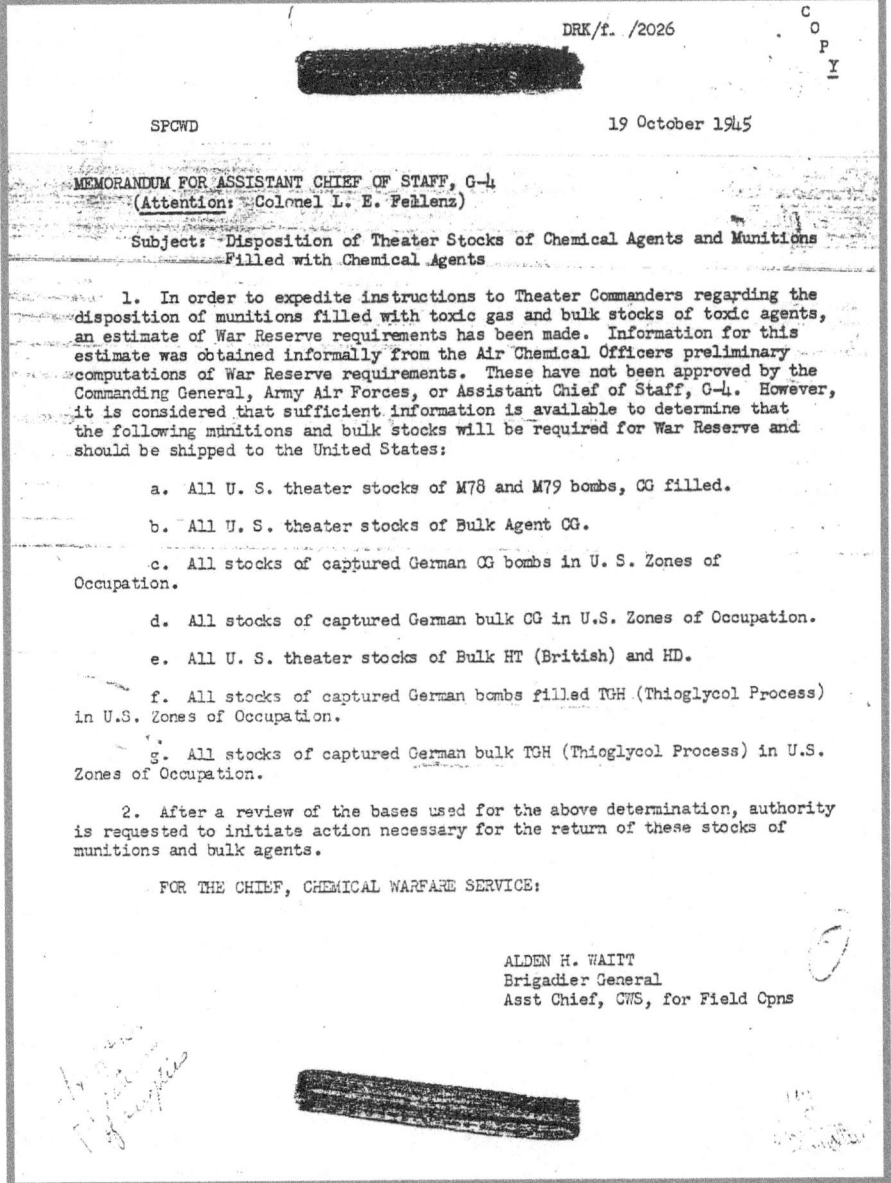

DRK/f. /2026

C
O
P
Y

SPCWD   19 October 1945

MEMORANDUM FOR ASSISTANT CHIEF OF STAFF, G-4
    (Attention: Colonel L. E. Fehlenz)

    Subject: Disposition of Theater Stocks of Chemical Agents and Munitions
              Filled with Chemical Agents

    1. In order to expedite instructions to Theater Commanders regarding the
disposition of munitions filled with toxic gas and bulk stocks of toxic agents,
an estimate of War Reserve requirements has been made. Information for this
estimate was obtained informally from the Air Chemical Officers preliminary
computations of War Reserve requirements. These have not been approved by the
Commanding General, Army Air Forces, or Assistant Chief of Staff, G-4. However,
it is considered that sufficient information is available to determine that
the following munitions and bulk stocks will be required for War Reserve and
should be shipped to the United States:

        a. All U. S. theater stocks of M78 and M79 bombs, CG filled.

        b. All U. S. theater stocks of Bulk Agent CG.

        c. All stocks of captured German CG bombs in U. S. Zones of
Occupation.

        d. All stocks of captured German bulk CG in U.S. Zones of Occupation.

        e. All U. S. theater stocks of Bulk HT (British) and HD.

        f. All stocks of captured German bombs filled TGH (Thioglycol Process)
in U.S. Zones of Occupation.

        g. All stocks of captured German bulk TGH (Thioglycol Process) in U.S.
Zones of Occupation.

    2. After a review of the bases used for the above determination, authority
is requested to initiate action necessary for the return of these stocks of
munitions and bulk agents.

    FOR THE CHIEF, CHEMICAL WARFARE SERVICE:

                            ALDEN H. WAITT
                            Brigadier General
                            Asst Chief, CWS, for Field Opns

Figure 16
Declassified CONFIDENTIAL 19 October 1945 Memorandum
Summarizing U.S. War Reserve Requirements

<u>30 Oct 45</u>

On 30 Oct 45, the War Department General Staff Logistics Group sent the following declassified SECRET message to the Commanding General, U.S. Forces, European Theater with an information copy to the Commanding General of U.S. Forces in Austria (Figure 17). This message is significant for at least two reasons:

It requests that further destruction of enemy war material in the European Theater – including France, Germany, and Austria – be deferred until new guidance is received from the Joint Chiefs of Staff. It appears from this message that the Secretary of State had intervened and requested the War Department to "restudy" its policy.

It implies that enemy war material was being destroyed not only in Germany, but in France and Austria.

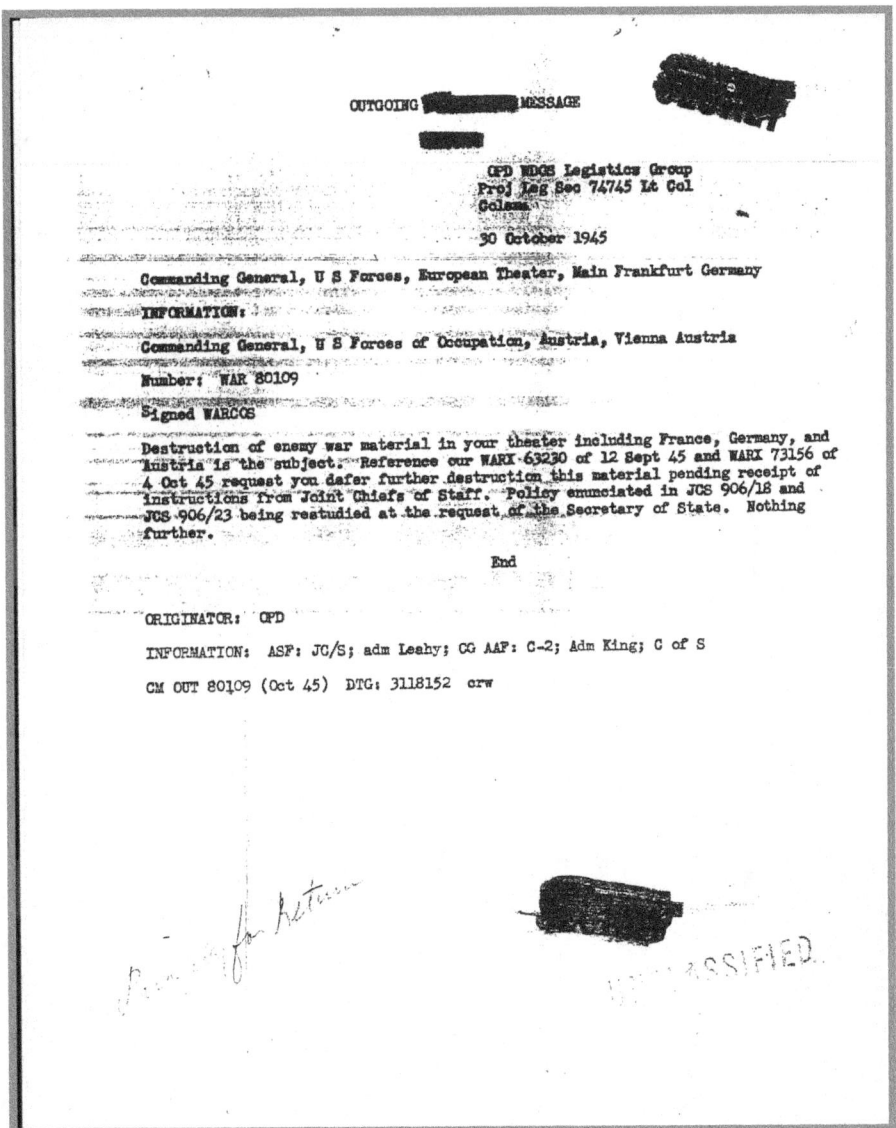

OUTGOING MESSAGE

OPD MDGS Logistics Group
Proj Log Sec 74745 Lt Col
Colem

30 October 1945

Commanding General, U S Forces, European Theater, Main Frankfurt Germany

INFORMATION:

Commanding General, U S Forces of Occupation, Austria, Vienna Austria

Number: WAR 80109

Signed WARCOS

Destruction of enemy war material in your theater including France, Germany, and Austria is the subject. Reference our WARX 63230 of 12 Sept 45 and WARX 73156 of 4 Oct 45 request you defer further destruction this material pending receipt of instructions from Joint Chiefs of Staff. Policy enunciated in JCS 906/18 and JCS 906/23 being restudied at the request of the Secretary of State. Nothing further.

End

ORIGINATOR: OPD

INFORMATION: ASF; JC/S; adm Leahy; CG AAF: C-2; Adm King; C of S

CM OUT 80109 (Oct 45) DTG: 3118152 crw

Figure 17
Declassified SECRET 30 October 1945 Message
Temporarily Halting the Destruction of
Enemy War Material in the European Theater –
including Germany, France, and Austria

<u>14 Dec 45</u>

In a 14 Dec 45 declassified CONFIDENTIAL Letter from Colonel John C. MacArthur, Director of Personnel, Plans, and Training to the Chief Chemical Warfare Officer, European Theater of Operations, the following summary of U.S. requirements for captured German CW was provided:

Requirements of CWS for Captured German Ammunition

(Letter 14 Dec 45)

ITEM	QUANTITY	FOR	REFERENCE
250 KG bombs Tabun filled	3,000	Technical	87411 (24 May 45)
105 MM artillery shell, Tabun filled	5,000	Technical	87411 (24 May 45)
Colored air craft smoke signal, red	200	C.W. Board	32692 (14 Jul 45)
Colored air craft smoke signal, violet	200	C.W. Board	32692 (14 Jul 45)
250 KG bombs filled thickened HKC 250 IIGB	100	C.W. Board	34527 (18 Jul 45)
250 KG bombs filled winterlost KC 250 GB	100	C.W. Board	34527 (18 Jul 45)
250 KG bombs filled CG	100	C.W. Board	34527 (18 Jul 45)
250 KG bombs filled LE-100 (Tabun)	100	C.W. Board	34527 (18 Jul 45)
Bombs KC 250 11 GR filled CG	42,400 or equivalent in bulk	War Reserve	83030 (14 Nov 45)
Bombs KC 250 GB filled, thiodiglycol mustard (naught) or Sommerlost)	451,700 or equivalent in bulk	War Reserve	83030 (14 Nov 45)
Rocket, smoke, 15 CM., WGR 41	100	C.W. Board	22391 (20 Nov 45)
Rocket, gelbring, 15 CM, WGR 41	100	C.W. Board	22391 (20 Nov 45)
Rocket, Grunring-Gelb, 15 CM., WGR 41, WKH	100	C.W. Board	22391 (20 Nov 45)
Rocket, Grunring 1, 15 CM., WGR 41, WKH	100	C.W. Board	22391 (20 Nov 45)
Rocket, 15 CM, nebelwerfer 41	12	C.W. Board	22391 (20 Nov 45)
250 Kg. bombs, Tabun filled	2,000		(Ltr 14 Dec 45)

Figure 18
Declassified CONFIDENTIAL 14 December 1945 Letter
Summary of U.S. Requirements for Captured German CW
as of 14 Dec 45

Apr 46

In April 1946, ten hulks were finally transferred to the War Department from the War Shipping Administration and made available to the Theatre for filling with ammunition and scuttling. The work of dumping, scuttling, and destruction was accelerated. Plans called for the disposal of all captured enemy ammunition in one year, but required an increase in personnel, both Army and civilian.[47]

28 May 46

The evolution of the U.S. plan now nearing completion, representatives of the Office of Military Government for Germany – United States (OMGUS) and Theater Headquarters met with German officials on May 28, 1946 to plan the turnover of all remaining stocks of captured enemy ammunition to OMGUS for demilitarization and salvage of component parts for use in the German economy. The plan provided for the transfer of approximately 250,000 tons of non-toxics and 70,000 tons of toxics for breakdown into component raw materials for use in the German economy in connection with the agriculture and manufacturing program.[48]

Jun 46

By June 1946, all U.S. requirements for CW from captured German stocks had been shipped, destruction of CW at depots in situ was underway, and the loading of hulks had started.[49]

1 Jul 46

On July 1, 1946, the first U.S. sea disposal convoy of Operation Davey Jones Locker was scuttled, nine months after the first British chemical weapons convoy. "Operation Davey Jones Locker" began in June 1946 and ran through August 1948. The actual outloading for all scuttling operations occurred at the Midgard Docks in Nordenham, Germany under the jurisdiction of the Bremerhaven Port Chemical Officer. Eleven ships were scuttled in this operation: nine ships in the Skaggerak Strait[50] and two hulks in the North Sea.

As sea disposal operations were being conducted by the Western powers, the U.S.S.R. mounted its own operation in what was to become the German Democratic Republic. Trainloads of CW were dispatched from storage sites and manufacturing plants mainly to the Baltic ports of Peenemünde and Wolgast. There the chemical weapons were loaded on barges or, in some cases, hulks. The U.S.S.R. chose an area northeast of Bornholm, Denmark, as its main dumping ground. The Gotland Basin, southeast of the Swedish island of Gotland, was also used as a Soviet dumping site. The barges were towed to sea and the munitions were dumped over the side, sometimes en route to the dumping site.[51]

There is no evidence of any agreement between the United States and the Soviet Union governing the German chemical weapons that were dumped into the sea between 1944 and 1948. Three government organizations which may have had some dealings with the Soviet Union regarding disposal operations, however, include:

- The Bipartite Control Office, OMGUS (Office Military Government for Germany – United States), which issued a certificate of clearance concerning the concluding phase of the disposal of captured enemy toxic materiel,
- The Bizonal Economic Council, which authorized a German corporation to set up a field plant at the St. Georgen depot to reduce certain CW stocks into commercially useable components, and
- The Continental Dumping Committee.[52]

According to the 1993 National Report of the Russian Federation entitled "Complex Analysis of the Hazard Related to the Captured German Chemical Weapons Dumped in the Baltic Sea", a total of approximately 70,500 tons (gross weight) of CW were found in the Russian Zone and that 12,035 tons (net weight) of chemical agents were sea disposed. The report also provides the final CW disposition strategy: "Destruction of the captured German CW was performed under the control of U.S.S.R. by the following means:

- Burning in the sites of storage in dedicated furnaces and in open pits,
- Exploding the artillery shells and aircraft bombs on the especially prepared grounds,
- Use of war chemical agents and their components in peaceful chemical industries,
- Dumping in sea.

Most of the bulk amount of German CW (up to 60%) was dumped in the sea."[53]

Given the reported (as opposed to actual) total of 70,500 tons, 60% would account for approximately 42,300 tons of CW being dumped into the Baltic. Estimating that chemical agents account for 10.0-32.5% of the gross weight of the munitions, this would equate to between 4,230 and 13,747 tons of chemical agents being disposed. Because a significant amount of CW disposed was in large containers, however, the reported net weight figure of 12,035 tons is reasonably corroborated.

Although no information was found regarding the evolution of a French plan for disposition of captured chemical weapons, one CW depot was reportedly found in the French Zone containing a total of approximately 8,000 tons of CW. A SECRET letter (now declassified), dated February 27, 1946, from HQ, BAOR, to the War Office provides some key data, however, regarding inventories and final disposition:

27 Feb 46

"1. The following information concerning stocks of German CW ammunition held by the French has been communicated to this Headquarters by Lt. Col. D'Anselme, Director Generale du Controle da Desarmament, Commandement en Chef Francais en Allemagne, Baden-Baden.

2. A total quantity of 8,000 tons of CW ammunition is held by the French at the Urlau Depot (near to Constance), this being made up as follows:

(a) 2,000 tons, shell charged "DM" (Adamsite),
(b) 3,000 tons, shell charged "H" (Mustard), and
(c) 3,000 tons, shell charged "GA" (Tabun).

3. Items in para 2(a) and (b) are being destroyed by the French, while assistance is being sought from this HQ (HQ, BAOR) with regard to the destruction of the Tabun (sub-para 2(c)) by deep sea dumping.[54]

<u>16 Jul 46</u>

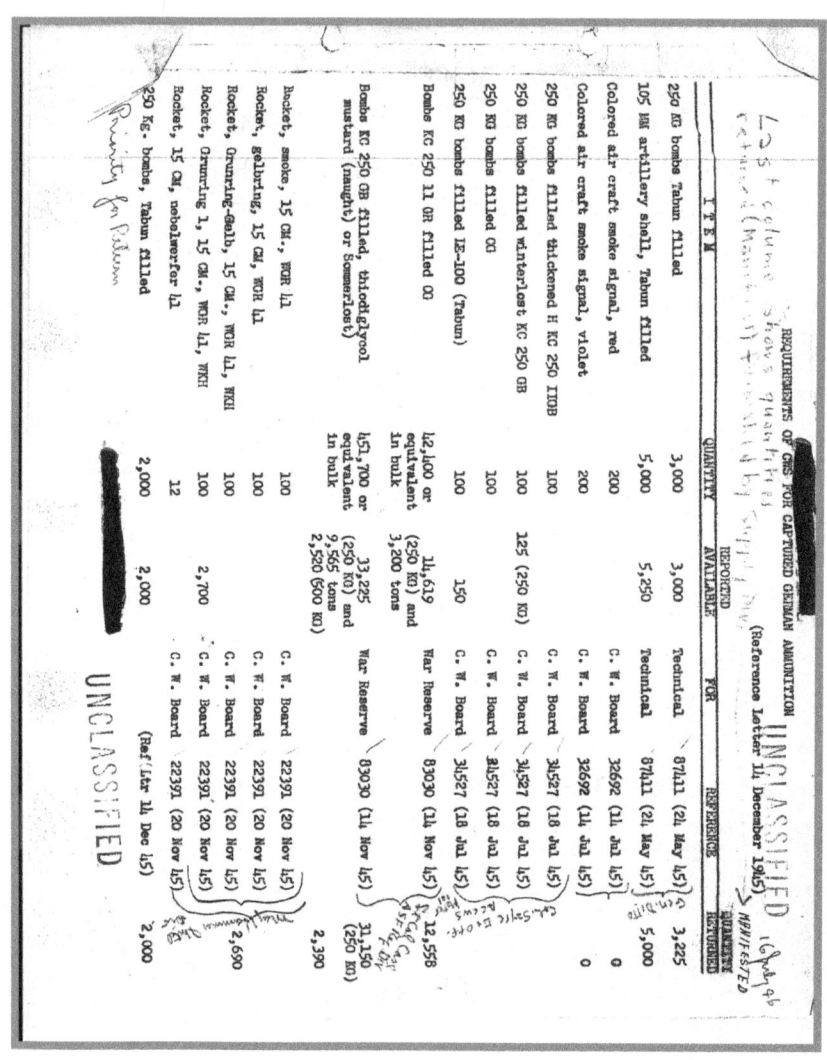

ITEM	QUANTITY	REPORTED AVAILABLE	FOR	REFERENCE	QUANTITY RETURNED
250 KG bombs Tabun filled	3,000	3,000	Technical	87411 (24 May 45)	3,225
105 MM artillery shell, Tabun filled	5,000	5,250	Technical	87411 (24 May 45)	5,000
Colored air craft smoke signal, red	200		C.W. Board	32692 (14 Jul 45)	0
Colored air craft smoke signal, violet	200		C.W. Board	32692 (14 Jul 45)	0
250 KG bombs filled thickened H KC 250 IKB	100		C.W. Board	34527 (18 Jul 45)	
250 KG bombs filled winterlost KC 250 GB	100	125 (250 KG)	C.W. Board	34527 (18 Jul 45)	
250 KG bombs filled GA	100		C.W. Board	34527 (18 Jul 45)	
250 KG bombs filled IE-100 (Tabun)	100	150	C.W. Board	34527 (18 Jul 45)	
Bombs KC 250 11 GR filled GA	100	14,619 (250 KG) and 3,200 tons	War Reserve	83030 (14 Nov 45)	12,558
Bombs KC 250 GR filled, thiodiglycol mustard (naught) or Sommerlost)	42,400 or equivalent in bulk	45,700 or equivalent in bulk 33,225 (250 KG) and 9,565 tons 2,520 (500 KG)	War Reserve	83030 (14 Nov 45)	31,150 (250 KG)
Rocket, smoke, 15 CM., WGR 41	100		C.W. Board	22391 (20 Nov 45)	2,390
Rocket, gelbring, 15 CM, WGR 41	100		C.W. Board	22391 (20 Nov 45)	
Rocket, Grunring-Gelb, 15 CM., WGR 41, WKH	100		C.W. Board	22391 (20 Nov 45)	
Rocket, Grunring 1, 15 CM., WGR 41, WKH	100	2,700	C.W. Board	22391 (20 Nov 45)	2,690
Rocket, 15 CM, nebelwerfer 41	12		C.W. Board	22391 (20 Nov 45)	
250 Kg. bombs, Tabun filled	2,000	2,000		(Ref Ltr 14 Dec 45)	2,000

UNCLASSIFIED

Figure 19
Declassified CONFIDENTIAL 16 July 1946 Summary of U.S. Requirements for Captured German CW

1.2 Chronology of Key Events in the American and British Zones of Occupation and US, UK, Russian and German Sea Disposal Operations

The following is a summary of the major CW-related events which occurred in both the American and British zones of occupation and US, UK, and Russian sea disposal operations which occurred from 1946 through 1990. The time span for the American Zone is the 1,244-day period (3 years and 5 months) between the occupation of the first chemical depot (Frankenberg) on March 29, 1945 and the scuttling of the final convoy [Davey Jones Locker (DJL) 10] on August 24, 1948. For the British Zone, the time span is the 1,830-day period (5 years) between the first meeting of the Standing Committee on War Material (SCWM) on December 4, 1944 and the final meeting of the Control Commission for Germany (British Element) Demilitarization Committee [CCG(BE)DC] on December 8, 1949. Refer to Appendix C for a detailed chronology of events.

1944

6 June	D-Day Invasion
12 September	The London Protocol
4 December	Standing Committee on War Material (SCWM) Meeting #1
9 December	SCWM Meeting #2
14 December	SCWM Meeting #3
20 December	SCWM Meeting #4

<u>1945</u>   (3 UK Skagerrak convoys with 17 total ships, 5 UK Atlantic operations with 5 total ships)

3 January	SCWM Meeting #5
10 January	SCWM Meeting #6
24 January	SCWM Meeting #7
4-11 February	Yalta Conference
7 February	SCWM Meeting #8
14 February	SCWM Special Meeting
21 February	SCWM Meeting #9
7 March	SCWM Meeting #10
21 March	SCWM Meeting #11
29 March	Frankenberg Chemical Depot occupied (1st of 5 U.S. Depots.)
4 April	SCWM Meeting #12
6 April	Wildflecken Chemical Depot occupied (2nd of 5 U.S. Depots)
17 April	S.S. Bantom and Cape Borda in U.K. waters with 958 tons of 150 lb bombs from the U.S. Army to be ultimately dumped in Beaufort's Dyke
18 April	SCWM Meeting #13
19 April	Grafenwöhr Chemical Depot occupied (3rd of 5 U.S. Depots)
25 April	U.K. MoD ammunition dumping experiment in St. Catherine's Deep

29 April	Schierling Chemical Depot occupied (4th of 5 U.S. Depots)
2 May	SCWM Meeting #14
3 May	St. Georgen Chemical Depot occupied (5th of 5 U.S. Depots)
	German Disposal Operations: In the last days of the war, two ships with 69,000 tabun shells and 5,000 tons of other chemical munitions were disposed in the Baltic's Little Belt.
8 May	V-E Day (Victory in Europe Day)
11 May	SCWM Meeting #15
8 June	SCWM Meeting #16
25 June	Chemical Munitions Sub-Committee (CMSC) Meeting #1
2 July	1st of 24 UK Atlantic Disposal Operations (1 ship: Empire Fal)
11 July	CMSC Meeting #2
15 July	Continental Ammunition Dumping Committee (CADC) Meeting #1
17 July-2 August	Potsdam Conference
1 August	CADC Meeting #2
7 August	SCWM Meeting #17
16 August	SCWM Meeting #18
16 August	CADC Meeting #3 (Details of CW dumping programme)

29 August	CADC Meeting #4 (Details of CW loading programme)
3 September	SCWM Meeting #19
11 September	2nd of 24 UK Atlantic Sea Disposal Operations (1 ship: Empire Simba)
17 September	SCWM Meeting #20
17 September	CADC Meeting #5
1 October	3rd of 24 UK Atlantic Sea Disposal Operations (1 ship: Empire Cormorant)
4 October	SCWM Meeting #21
4 October	Convoy CW1 (1st UK Skagerrak CW Convoy, 5 Ships)
5 October	CADC Meeting #6
17 October	Convoy CW2 (2nd UK Skagerrak CW Convoy, 7 Ships)
18 October	SCWM Meeting #22
30 October	4th of 24 UK Atlantic Sea Disposal Operations (1 ship: Wairuna)
15 November	SCWM Meeting #23
17 November	Convoy CW3 (3rd UK Skagerrak CW Convoy, 5 Ships)
6 December	SCWM Meeting #24
7 December	CADC Meeting #7
20 December	SCWM Meeting #25

30 December	5th of 24 UK Atlantic Sea Disposal Operations (1 ship: Botlea)
1945-1946	The U.S. disposed of "unspecified amounts" of CW in the Mediterranean Sea off the west coast of Italy near Naples and off the east coast near both Bari and Brindisi.
1946	(4 US Convoys with 5 total ships, 5 UK Skagerrak convoys with 13 total ships, 5 UK Atlantic operations with 5 total ships)
10 January	SCWM Meeting #26
21 January	CADC Subcommittee Special Meeting
24 January	SCWM Meeting #27
16 March	Convoy CW4 (4th UK Skagerrak CW Convoy, 4 Ships)
18 March	CADC Meeting #8
18 March-18 April	Mustard filling project at St. Georgen Chemical Depot
21 March	SCWM Meeting #28 (Final meeting of the SCWM)
27 April-3 July	Mustard burning project at St. Georgen Chemical Depot
1 May	Shipments of CW to the port of Bremerhaven for scuttling and to Antwerp for shipment back to the U.S. had started
28 May	Representatives from the Office of Military Government for Germany –

U.S. (OMGUS) and Theatre HQ met with German officials to plan turning over all remaining captured enemy ammunition to OMGUS for demilitarization and salvage of component parts for use in the German economy.

June	All U.S. requirements for CW from captured German stocks had been shipped, destruction of CW at depots in situ was underway, and initial sinkings of loaded hulks had started.[55]
24 June	All scuttling operations and in situ destruction halted on order from HQ, U.S. Forces, European Theater (USFET). Final destruction responsibility transferred from USFET/Continental Base Section to OMGUS.
1 July	Convoy DJL1 – First U.S. Convoy in "Operation Davey Jones Locker"
2 July	Convoy DJL1 completed with the scuttling of the UJ-305.
13 July	Convoy CW5 (5th UK Skagerrak CW Convoy, 2 Ships)
14 July	Convoy DJL2
July-October	The U.S. disposed of 1,700 tons of mustard and lewisite-filled bombs off the coast of St. Raphael, France

25 August	6th of 24 UK Atlantic Sea Disposal Operations (1 ship: Empire Peacock)
29 August	Control Group for Germany (British Element) Demilitarization Committee [CCG(BE)DC] Meeting #1
30 August	Convoy DJL3
3 September	7th of 24 UK Atlantic Sea Disposal Operations (1 ship: Empire Nutfield)
8 September	Convoy CW6 (6th UK Skagerrak CW Convoy, 2 Ships)
26 September	CCG(BE)DC Meeting #2
October	All remaining toxics in the U.S. Zone began to be consolidated at St. Georgen
1 October	8th of 24 UK Atlantic Sea Disposal Operations (1 ship: Kindersley)
12 October	Convoy CW7 (7th UK Skagerrak CW Convoy, 3 Ships)
31 October	CCG(BE)DC Meeting #3
2 November	9th of 24 UK Atlantic Sea Disposal Operations (1 ship: Empire Woodlark)
11 November	10th of 24 UK Atlantic Sea Disposal Operations (1 ship: Lanark)
15 November	CCG(BE)DC Meeting #4
12 December	CCG(BE)DC Meeting #5
21 December	Convoy CW8 (8th UK Skagerrak CW Convoy, 2 Ships)

1946-1978	Russian loose-dump disposal operations conducted in the Baltic Sea
1947	(4 US Convoys with 4 total ships, 2 UK Skagerrak Convoys with 2 total ships, 5 UK Atlantic operations with 5 total ships)
23 January	CCG(BE)DC Meeting #6
5 February	11th of 24 UK Atlantic Sea Disposal Operations (1 ship: Dora Oldendorf)
27 February	CCG(BE)DC Meeting #7
15 March	Wildflecken Chemical Depot closed
15 March	Grafenwöhr Chemical Depot closed
15 March	European Command (EUCOM) replaced U.S. Forces European Theater (USFET)[56]
27 March	CCG(BE)DC Meeting #8
April	Additional sea disposal operations proposed by United States European Command (EUCOM) to the Office of Military Government for Germany – U.S. (OMGUS)
1 April	All remaining toxics in the U.S. Zone were consolidated at St. Georgen
10 April	Schierling Chemical Depot closed
10 April	Frankenberg Chemical Depot closed
15 April	Inventory of all remaining CW consolidated at St. Georgen
9 May	CCG(BE)DC Meeting #9

12 May	U.S. Rail shipments from St. Georgen to the port of Bremerhaven started
17 May	Convoy CW9 (9th UK Skagerrak CW Convoy, 1 ship)
1 June	St. Georgen Chemical Depot closed
6 June	Convoy DJL4
6 June	Convoy CW10 (Final UK Skagerrak CW Convoy, 1 Ship)
19 June	CCG(BE)DC Meeting #10
20 June	Convoy DJL5
30 June	Convoy DJL6
18 July	Convoy DJL7
24 July	CCG(BE)DC Meeting #11
27 July	12th of 24 UK Atlantic Sea Disposal Operations (1 ship: Empire Lark)
9 August	13th of 24 UK Atlantic Sea Disposal Operations (1 ship: Leighton)
8 September	14th of 24 UK Atlantic Sea Disposal Operations (1 ship: Thorpe Bay)
12 September	CCG(BE)DC Meeting #12
17 October	CCG(BE)DC Meeting #13
3 November	15th of 24 UK Atlantic Sea Disposal Operations (1 ship: Margo)
13 November	CCG(BE)DC Meeting #14

18 December	CCG(BE)DC Meeting #15 (the 32nd and final meeting occurred on 8 Dec 49)
December	Remaining quantity of CW after CW10 disposed of by burning
December	Final BAOR Disarmament Progress Report

1948 (2 US Convoys with 2 total ships, 3 UK Atlantic operations with 3 total ships)

29 January	CCG(BE)DC Meeting #16
1 March	16th of 24 UK Atlantic Sea Disposal Operations (1 ship: Harm Freitzen)
24 July	Convoy DJL8
22 August	17th of 24 UK Atlantic Sea Disposal Operations (1 ship: Empire Success)
24 August	Convoy DJL9 – Final U.S. Convoy of Captured CW
22 September	18th of 24 UK Atlantic Sea Disposal Operations (1 ship: Miervaldis)
25 November	CCG(BE)DC Meeting #24

1949 (1 UK Atlantic operation with 1 ship)

13 January	CCG(BE)DC Meeting #25
20 June	19th of 24 UK Atlantic Sea Disposal Operations (1 ship: Empire Connyngham)

| 8 December | CCG(BE)DC Meeting #32 (Final meeting) |

1955 (1 UK Atlantic Operation with 1 ship)

| 27 July | 20th of 24 UK Atlantic Sea Disposal Operations (1 ship: Empire Claire) (Operation Sandcastle) |

1956 (4 UK Atlantic Operations with 4 ships)

| 30 May | 21st of 24 UK Atlantic Sea Disposal Operations (1 ship: Vogtland) (Operation Sandcastle) |

| 23 July | 22nd of 24 UK Atlantic Sea Disposal Operations (1 ship: Kotka) (Operation Sandcastle) |

| Jun-Sep | 23rd and 24th of 24 UK Atlantic Sea Disposal Operations (2 ships: Unknown) |

1960

| March | German Disposal Operations: The two ships laden with 69,000 tabun shells which were scuttled in the Baltic's Little Belt before the end of the war were surfaced. The 69,000 nerve gas shells were encapsulated in concrete and re-dumped in the Bay of Biscay. |

| **1989-90** | Suspected Russian disposal operations in the Baltic |

CHAPTER 2

ACCOUNTING OF ALL CAPTURED CHEMICAL WEAPONS

2.1 Chemical Warfare Agents Produced by the Third Reich[57], 1935-1945

Soon after the end of World War I the German military became convinced that the 1919 Treaty of Versailles was so humiliating for Germany that another war was inevitable. A so-called 'struggle for liberation' was expected to start in the years just after the war. German military strength had been greatly weakened and it would not have been possible to increase Germany's conventional weapon capability rapidly. Instead, chemical warfare agents (CWA) were considered as a possible way to compensate for the lack of conventional capability. Those who favored the development of chemical warfare agents expressed the belief that these agents were 'natural' and 'superior' weapons as the use of them required both discipline and intelligence. Experiences in World War I had also demonstrated the effectiveness of chemical weapons against soldiers in trenches. Preparations for resumption of the production of chemical warfare agents began as early as 1923.

Prior to the assumption of power by the National Socialist Party in 1933, the quantity of chemical warfare agents produced in Germany was negligible. However, when the chemical industry became interested in increasing its production, the situation changed. In March 1935, the stockpile of CWA was only 300 tons of chloroacetophenone (CN) and 700 tons of thiodiglycol, a precursor used in the production of mustard gas (H). An additional 300 tons of thiodiglycol were in production and industry possessed the capability of manufacturing the chemical warfare agents Clark I (DA), Clark II (DC), and Diphosgene (DP). By 1936, two plants were producing chemical warfare agents (thiodiglycol[58] and arsinöl[59]) and one plant was filling mustard gas munitions.[60]

Although plans existed for a robust CW production program, progress was impeded primarily because of a general shortage of raw material. Iron, for example, which had originally been allocated for mustard gas storage tanks, was confiscated by the government to be used for other purposes. In 1939 the stockpile of mustard gas (H) was only 8,000 tons.

As events of the war escalated, shortages of basic raw material worsened in addition to shortages of chlorine, phosgene, sodium cyanide, ethylene, arsenic, and other chemicals. Recognizing the difficulties inherent in attempting to meet the goals of the CW production program, on October 1, 1943 an emergency program was inaugurated to limit production to a few types of chemical warfare agents. This meant that by the end of 1943 production plants had to abandon many of their specialized pursuits and postpone the purchase of new equipment. Production was also negatively affected by the fact that the branches of the armed forces – the army, air force, and navy – did not remove their stocks of these agents from the storage depots at the production plants, which thus remained full. Additionally, there was disagreement among the branches of the armed forces about the importance of various chemical agents.

By the spring of 1945 it was apparent that Germany would lose the war and the problem of how to handle the munitions filled with chemical agents became acute. All of the storage depots were expected soon to fall under the control of the advancing Allied forces[61] – which they did. The first CW depot in the American zone was taken over on March 29, 1945. By July 15, 1945, the Continental Ammunition Dumping Committee (CADC) was already prepared to execute the dumping program for all CW captured in the British Zone.

Figure 20 provides an accounting of all chemical warfare agents produced in Germany between 1935 and 1945 by production facility location[62], including planned capacity in tons per month. Two primary source documents were used to corroborate data: "German Munition Plants and Depots in Germany during World War II" published by the Historical Division, U.S. Army Chemical and Biological Defense Command and "The Challenge of Old Chemical Munitions and Toxic Armament Wastes" published by the Stockholm International Peace Research Institute. Figure 21 provides total production tonnage by chemical warfare agent.

Although reports such as "Dumped Chemical Weapons in the Sea – Options" published in 1999 by the Dr. A. H. Heineken Foundation for the Environment, The Netherlands, state "before and during WWII, the German chemical warfare industry produced and accumulated 65,000 tons (net weight) of mustard gas, chloroacetophenone, and different arsenic-containing compounds"[63], current research indicates a total of approximately 79,889 tons of CWA were produced in Germany between 1935 and 1945.

PRODUCTION FACILITY	CHEMICAL AGENT	PLANNED CAPACITY (TONS/MONTH)	TOTAL PRODUCTION (TONS)
Ammendorf	Nitrogen Mustard (HN)	50	1,928
	Sulfur Mustard (HD)	700	24,097
Berlin Haselhorst	Clark I (DA)	90	1,500
	Clark II (DC)	-	100
Dyhernfurth	Tabun (GA)	1,000	12,000
	Cyanogen Chloride (CK)	20	20
Falkenhagen	N-Stoff/Chlorotriflouride	10	50
Gendorf	Sulfur Mustard (HD)	1,500	3,500
Hüls	Sulfur Mustard (HD)	600	250
Ludwigshafen	Chloroacetophenone (CN)	70	3,133
Munster	Sarin (GB)	-	0.5
	Excelsior (arsenical)	-	10
Seelze/Hannover	Chloroacetophenone (CN)	140	4,000
Stassfurt	Arsine Oil	270	7,500
	Clark I (DA)	-	12,000
Uerdingen	Adamsite (DM)	200	3,900
Wolfen Bitterfeld	Phosgene (CG)	400	5,900
		TOTAL:	79,889

Figure 20
Total Chemical Warfare Agents Produced by the Third Reich,
1935-1945
By Production Facility

CHEMICAL AGENT	TOTAL TONNAGE
Arsinöl /Arsine Oil	7,500
Phosgene (CG)	5,900
Cyanogen Chloride (CK)	20
Chloroacetophenone (CN)	7,133
Clark I (DA)	13,500
Clark II (DC)	100
Adamsite (DM)	3,900
Excelsior (arsenical)	10
Tabun (GA)	12,000
Sarin (GB)	0.5
Sulfur Mustard (HD)	27,847
Nitrogen Mustard (HN)	1,928
N-Stoff / Chlorotriflouride	50
TOTAL:	79,889

Figure 21
Total Chemical Warfare Agents
Produced by The Third Reich, 1935-1945
by Chemical Agent

2.2 Characteristics of CW Produced by the Third Reich

Munitions filled with chemical warfare agents were marked with colored crosses using an identification system implemented during World War I. This system aimed to provide a simple means of differentiating these shells from conventional shells. The following markings were used: a white cross for tear gas agents (lachrymators), a blue cross for vomiting agents (including arsenicals such as Clark I and Clark II), a yellow cross for blistering agents (including vesicants such as sulphur mustard), and a green cross for choking agents (including asphyxiates such as phosgene and diphosgene). During the Third Reich, the armed forces continued to categorize munitions according to this marking system to some extent, but munitions were no longer marked using the system of color-coded crosses.

During World War II, German munitions filled with CWA were marked using colored rings, letters, and numbers. The ring color indicated the primary physiological effect of the CWA that would be released on detonation and did not necessarily identify its explosive charge or chemical fill. Blue (Blau - Bu) indicated a sternutator[64], green (Grun – Gr) indicated a choking agent or one which had a systemic effect, yellow (Gelb – Gb) indicated a vesicant[65], and white denoted a lachrymator.[66] Numbers were used to distinguish the various charges which had similar physiological effects but different levels of effectiveness. Letters indicated the exact nature of the chemical fill.[67]

To facilitate an understanding of the significance of the total tonnage of CW found in Germany by the Allies, Figures 22-25 provide consolidated reference information regarding munition ring colors, numbers, and letters and the chemical warfare agents used.

NOMENCLATURE	GERMAN NAME	CHEMICAL AGENT, NATO/US CODE, AND BURSTER
Blue Ring 1	Blauring 1	Adamsite (DM), Exterior Burster
Blue Ring 2	Blauring 2	Adamsite (DM) and Arsinöl, Central Burster
Blue Ring 3	Blauring 3	Adamsite (DM), Base Ejection Generator
One Green Ring	Grünring	Mustard Gas (H), Medium Burster
Green Ring 1	Grünring 1	Nitrogen Mustard (HN), Large Burster
Green Ring 2	Grünring 2	Phosgene (CG) or Diphosgene (DP)
Green Ring 3	Grünring 3	Tabun (GA), Head Burster
Green Ring-Yellow	Grünring-Gelb	Mustard Gas (H), Large Burster
One Yellow Ring	Gelbring	Mustard Gas (H), Small Burster
Double Yellow Ring		Thickened Mustard Gas (HT), Small Burster
One White Ring	Weissring	Chloroacetophenone (CN)

Figure 22
Color-Coding System for German CW Munitions
Produced during the Third Reich

Between 1935 and 1945, an imposing number of organo-arsenic and mustard compounds were developed and tested for use as chemical warfare agents. For example, components were intentionally added to the original mustard products to reduce the freezing point for winter use or to improve persistency on the ground and on affected surfaces of military equipment after CW agent dispersal. One such formulation is the viscous mustard Zählost (see Figure 23). Besides the powerful enhancement of viscosity, the thickening materials (e.g., chlorinated rubber, waxes, and polystyrene) transformed the chemical agent formulations to become completely insoluble in usual and technically available solvents. Viscous mustard also posed considerable technical disposal problems. Significant amounts of arsine oil were used to winter-proof various types of mustard gas.

The following table lists German chemical warfare agents by letter code, German name, NATO/US code, and typical formulations in cases where CWA were combined.

LETTER CODE	GERMAN NAME	CHEMICAL AGENT	NATO/ US CODE	FORMULATION
A	Chloraceto-phenon	Chloroaceto-phenone	CN	
A-Ol	Arsinöl	Arsine Oil		50% Pfifficus, 35% Clark I, 5% Triphenylarsine, 5% Arsenic Chloride
B		Thiodiglycol Mustard and Arsinol		50% Thiodiglycol Mustard, 50% Arsinol
C		Thiodiglycol Mustard and Chlorobenzene		50% Thiodiglycol Mustard, 50% Chlorobenzene
C I	Clark I	Clark I	DA	Diphenylchloroarsine
C II	Clark II	Clark II	DC	Diphenylcyanoarsine

D	Direklost	Thickened Mustard made from "B"		
F	Phosgen	Phosgene	CG	
G	Tabun	Tabun	GA	
GA	Tabun	Tabun and Chlorobenzene		
H	Diphosgen	Diphosgene	DP	
K	N-Lost	Nitrogen Mustard	HN	
L		Thiodiglycol Mustard and Anthracene Oil		67% Thiodiglycol Mustard, 33% Anthracene Oil
M	Adamsit	Adamsite	DM	
N		Clark I and Arsinöl		40% Clark I, 60% Arsinöl
O	S-Lost (Sommerlost) or Oxollost	Sulfur Mustard / Thiodiglycol Mustard (Summer Mustard)	HD	
P		Hydrogen Cyanide	AC	
Z-OA	Zählost[68]	Viscous Mustard		See footnote
Z-OM	Zählost	Viscous Mustard		45% Sulfur Mustard, 36% Propyl Mustard, 9% Oxygen Mustard, 7% Polystyrene, 3% S-Wax

Figure 23
German Chemical Warfare Agents Listed by German Letter
Code Designation

German chemical warfare agents are indexed here by German name, including NATO/US code, and formulation, in cases where CWA are combined.

GERMAN NAME	CHEMICAL AGENT	NATO/US CODE	FORMULATION
Adamsit	Adamsite	DM	
Arsinöl	Arsine Oil		50% Pfiffikus, 35% Clark I, 5% Triphenylarsine, 5% Arsenic Chloride
Chloroacetophenon Or CN-Stoff	Chloroacetophenone	CN	
Clark I	Clark I	DA	Diphenylchloroarsine
Clark II	Clark II	DC	Diphenylcyanoarsine
Dick	Ethyldichloroarsine	ED	
Diphosgen or Perstoff	Diphosgene	DP	
Direktlost	Thickened Mustard made from "B"		
Methyldick or Medikus	Methyldichloroarsine	MD	
N-Lost	Nitrogen Mustard	HN	
Pfiffikus	Phenyldichloroarsine	PD	
Phosgen	Phosgene	CG	
S-Lost (Sommerlost) or Oxollost	Sulfur Mustard / Thiodiglycol Mustard (Summer Mustard)	HD	
Tabun	Tabun	GA	
Winterlost	Mustard	H	Sulfur mustard mixed with phenyldichloroarsine (pfifficus) to achieve a lower freezing point than pure mustard
Zählost	Viscous Mustard		See Figure 23 for formulations

Figure 24
German Chemical Warfare Agents
Listed by German Name

According to the study conducted by the Stockholm International Peace Research Institute (SIPRI) in 1997, approximately 7,974,350 individual pieces of charged land-based CW, such as artillery rounds, rockets, and mortar bombs, were captured by the Allies in 1945.[69] Artillery rounds alone accounted for 94% of the land-based CW. Figure 25 provides a sampling of key data on the artillery rounds, aerial bombs, cans, drums, and containers that were captured and ultimately sea-disposed. Individual munition weights and chemical filling weights are crucial data when estimating of total amount of chemical warfare agents sea-disposed based upon known total weights. These calculations will be addressed in detail in Chapter 3.

Principal sources of munition and chemical weights for artillery shells and aerial bombs include:

1) "Kampfstoff-Munition: Die Kampfstoffmunition des Zweiten Weltkrieges" (German chemical munitions handbook),

2) "Recovered Old Arsenical and Mustard Munitions in Germany: Technologies, Plans, and Problems", Wehrwissenschaftliches Institut für Schutztechnologien-ABC Schutz, and

3) "German Chemical Warfare Materiel" – a declassified report of the Intelligence Division, Chemical Warfare Service, HQ, European Theater of Operations, United States Army (ETOUSA).

Chemical weights for cans, drums, and containers were obtained from the 1993 National Report of the Russian Federation.

	CHEMICAL AGENT	TOTAL WEIGHT (KG)	CHEMICAL WEIGHT (KG)	% CHEMICAL
10.5 cm Shells				
Blue Ring 1	Adamsite (DM)	14.8	0.55	3.70%
Blue Ring 3	Adamsite/Nitrocellulose 50/50	14.1	0.837	5.90%
Green Ring 1	Nitrogen Mustard (HN)	14.2	0.9	6.30%
White Ring	Chloroacetophenone (CN)	14.5	1.219	8.40%
15 cm Shells				
Blue Ring 1	Adamsite (DM)	42.3	1.485	3.50%
Green Ring 1	Nitrogen Mustard (HN)	38.1	2.9	7.60%
Yellow Ring	Mustard (H)	37.4	4.3	11.50%
White Ring	Chloroacetophenone (CN)	42.5	3.5	8.20%
Aerial Bombs				
KC 50 II Bu	Adamsite (DM) and Nitrocellulose	43	14	32.60%
KC 250 Gb	Sulfur Mustard (HD)	160	100	62.50%
KC 250 II Gb	Thickened or Viscous Mustard	165	105	63.60%
KC 250 Gr	Sulfur Mustard (HD)	166	100	60.20%
KC 250 II Gr	Phosgene (CG)	160	100	62.50%
KC 250 III Gr	Tabun (GA)	149	93	62.40%
KC 250 W	Chloroacetophenone (CN)	140	100	71.40%
KC 500 II Gr	Phosgene (CG)	472	215	45.60%
KC 500 W	Chloroacetophenone (CN)	467	258	55.20%
Cans			0.3	
Drums			100	
Containers			1,000.00	

Figure 25
Characteristics of Principal Shells and Aerial Bombs
Captured and Sea-Disposed

2.3 Total CW Tonnage Found on German Territory by the Allies

Estimating the gross weight of all captured chemical weapons, the 1993 National Report of the Russian Federation, "Complex Analysis of the Hazard Related to the Captured German Chemical Weapons Dumped in the Baltic Sea," states that Germany possessed approximately 311,200 tons of CW at the end of World War II in the following Zonal distribution:

- American Zone 104,500 tons (including 10.5 tons in bulk)
- British Zone 126,700 tons (including 4.2 tons in bulk)
- Russian Zone 70,500 tons (including 8.0 tons in bulk)
- French Zone 9,500 tons (including 0.2 tons in bulk)

The 1994 Report to the 15th Meeting of the Helsinki Commission from the Ad-Hoc Working Group on Dumped Chemical Munition (HELCOM CHEMU), estimates a total of 296,103 tons:[70]

- American Zone 93,995 tons
- British Zone 122,508 tons
- Russian Zone 70,500 tons
- French Zone 9,100 tons

In 1999, Russian Major General Boris T. Surikov, in his synopsis "Dumped Chemical Weapons in the Sea – Options," estimated 296,253 tons.[71] Current research indicates, however, a total of 320,819 tons in the following distribution.

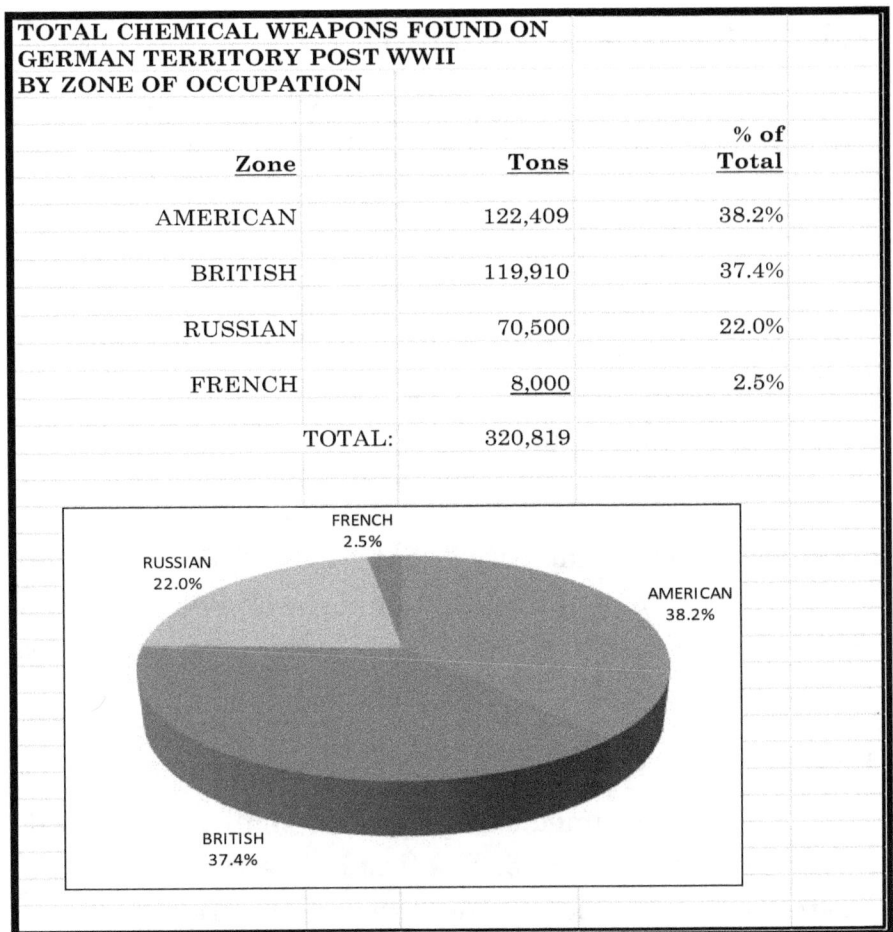

TOTAL CHEMICAL WEAPONS FOUND ON GERMAN TERRITORY POST WWII BY ZONE OF OCCUPATION		
Zone	Tons	% of Total
AMERICAN	122,409	38.2%
BRITISH	119,910	37.4%
RUSSIAN	70,500	22.0%
FRENCH	8,000	2.5%
TOTAL:	320,819	

Figure 26
Total CW Found on German Territory
by Zone of Occupation

Derivations of these findings are addressed in the following sections.

2.3.1 The American Zone

A total of 122,409 tons of chemical weapons were found in the American Zone. All captured chemical weapons were consolidated at five former German Army and Air Force CW depots:

- Frankenberg
- Wildflecken
- Grafenwöhr
- Schierling, and
- St. Georgen

U.S. records indicate that 2,801,905 individual chemical weapons, including mines, candles, sprayers, shells, rockets, and aerial bombs, and bulk containers were captured in these five depots.

Frankenberg Depot

U.S. forces assumed control of Frankenberg Chemical Depot on March 29, 1945. The first of five such depots located in the American Zone, Frankenberg remained operational for 742 days. On April 10, 1947 the depot was closed and the 227th Chemical Base Depot Company was deactivated.

Pictured below is the main entrance to Frankenberg Depot in December 1946. The sign on the right gate states: "Welcome to the Frankenberg CWS-CEM Depot. Speed Limit = 15 MPH. Military courtesy and discipline enforced. All visitors will receive a gas mask and will carry it while in the depot. Gas masks will be returned to the guard on departure."

Figure 27
Frankenberg Depot – Main Gate
December 1946

Figure 28
Frankenberg Depot – CW Storage
December 1946

Figure 29
Frankenberg Depot – Crated CW
December 1946

Figure 30
Frankenberg Depot – Chemical Bombs
December 1946

Wildflecken Depot

U.S. forces assumed control of Wildflecken Chemical Depot on April 6, 1945. The second of five such depots located in the American Zone, Wildflecken remained operational for 708 days. A significant amount of the chemical munitions found at Wildflecken in the early spring of 1945 were in disarray as evidenced by the photograph at Figure 31. Historians attribute this to the Allies' rapid advance. On March 15, 1947 the depot was closed and the 13th Chemical Maintenance Company was deactivated.

Figure 31
Wildflecken Depot – Chemical Munitions
December 1946

Figures 32a and 32b
Wildflecken Depot – CWS Supervision
December 1946

Figures 33 and 34
Wildflecken Depot – Rail Loading Operations
December 1946

Grafenwöhr Depot

U.S. forces assumed control of Grafenwöhr Chemical Depot on April 19, 1945. The third of five such depots located in the American Zone, Grafenwöhr remained operational for 695 days. Chemical officers from the Third U.S. Army conducted a preliminary survey of the Grafenwöhr storage area on April 24th and a preliminary inventory was conducted before January 1946. It was not until the 18th Chemical Maintenance Company arrived in February, 1946 that a detailed inventory was conducted. On March 15, 1947, the depot was closed and the 18th Chemical Maintenance Co. was deactivated.

There were no ammunition bunkers, warehouses, or storage buildings of any kind in the Grafenwöhr Depot when the Americans arrived. All ammunition was piled in the open, along roadways, and in the woods. The dump consisted of approximately 2,000 acres of wooded, rolling land located about 3/4 miles NE of Grafenwöhr on the road between that town and Pechof. No buildings other than two small one-room hunting cabins were in the area. No storage of decontamination or protective equipment was found and no German civilian or military personnel who were familiar with the storage could be located.

The munitions in storage consisted of approximately 2,000,000 rounds of artillery shells, Nebelwerfer rockets, landmines, and approximately 75% of the storage is of German manufacture and design; the other 25% consisted of captured French, Hungarian, and a small lot of Russian and Czechoslovakian shells. Calibers varied from 77 mm to 210mm, the greatest proportion being 105 mm. Dates of manufacture extended from 1918 (French phosgene shells) to February 1945.

Figures 35 and 36
Grafenwöhr Depot as Found by the American Forces and
View of the Main Road of the Depot Area, 21 May 1946

Figures 37 and 38
Grafenwöhr Depot – Repacking German 105mm Toxic
Artillery Shells and Field Stacks of 75mm Mustard Gas-Filled
Artillery Shells, 21 May 1946

Schierling Depot

U.S. forces assumed control of Schierling Chemical Depot on April 29, 1945. The fourth of five such depots located in the American Zone, Schierling remained operational for 711 days. On April 10, 1947, the depot was closed.

Figure 39
Schierling Depot – Security Check at the Main Gate
17 May 1946

Figures 40 and 41
Schierling Depot – Aerial views of Depot Service Buildings
and consolidated Tabun gas bombs, 17 May 1946

Figures 42 and 43
Schierling Depot – Field Stacks of 250kg Tabun bombs and
Earth Pit Disposal of a Tabun Leaker, 17 May 1946

St. Georgen Depot

U.S. forces assumed control of St. Georgen Chemical Depot on May 3, 1945. Despite being the last of five such depots to surrender to U.S. forces, St. Georgen remained operational longer than any other depot in the American Zone - 759 days. On June 1, 1947, the depot closed and the 193rd Chemical Depot Company, which had been in charge of depot activities since January 1946, was deactivated.

The mustard burning project at St. Georgen Chemical Depot was completed on 3 July 1946. 4,138 metric tons of captured German liquid mustard and 745 metric tons of Clark I and Clark II were disposed of by means of burning. All work was carried out by German civilian laborers under the close supervision of American officers and enlisted men. Activities were coordinated with local civilian authorities. Mustard was burned only on those days when wind would carry the smoke into the previously designated area and only when the wind was blowing not less than three miles or more than ten miles per hour. Winds and weather conditions were constantly checked and burning was halted whenever there was sufficient change in wind direction or velocity to carry the smoke outside of the previously designated limits. Considerable concern on the part of the German populace was felt at the beginning of the burning program which subsided as time went on and no evidence of any damage could be detected. As far as could be determined, there were no serious or lasting injuries to life or property.

Figure 44
St. Georgen Depot – Main Entrance
May 1946

Figure 45
St. Georgen Depot – Bulk Toxics
9 May 1946

View of the barrels of Hungarian and German mustard gas at the St. Georgen chemical warfare depot – the principal German center for processing and storage of toxic agents.

Figure 46
St. Georgen Depot – Field Stacks, 10 May 1946
Field stacks of mustard gas filled land mines

Figure 47
St. Georgen Depot – Outloading Operations
8 May 1946

View through a bunker entrance at St. Georgen Chemical Warfare Depot of workers loading railroad freight cars with Green 1 mustard-filled artillery shells. 70 bunkers were on the depot site containing up to 20,000 artillery shells.

Figure 48
St. Georgen Depot – Unfilled Toxic Shells
10 May 1946

Rows of 10.5mm German toxic artillery shell cases in a storage warehouse St. Georgen. This warehouse is connected with the filling plant by an elaborate underground conveyor system.

Figure 49
St. Georgen Depot – Mustard Burning Project, View 1
10 May 1946

A layer of chloride of lime was placed in the burning pit which when contacting the mustard gas, burst into flame and the liquid gas was completely destroyed.

Figure 50
St. Georgen Depot – Mustard Burning Project, View 2
8 May 1946

Smoke clouds, caused by burning mustard gas, containing arsenical particles. The prevailing winds were to the west, therefore east of the burning point, civilians were evacuated for a sector extending 3.5 km. On favorable days an average of 700 tons of mustard were burned in each of two pits at St. Georgen.

BOTTOM LINE: A total of 122,409 tons of chemical weapons were found in the American Zone. This number is reflected in the Total Tonnage Table. (Figure 55)

2.3.2 The British Zone

With specific regard to chemical munitions, the July 1945 report of the British Army of the Rhine (BAOR) noted that "it is believed that all major depots have now been discovered, the contents of the small dumps have been removed to the larger dumps (all of which are guarded), and there is a total of approximately 111,000 tons of enemy CW munitions in the British Zone."[72] All captured chemical weapons were consolidated at fourteen former German ammunition depots/facilities from an initial total of seventeen depots.

On 23 August 1945, the Commander-in-Chief, 21st Army Group transmitted an inventory of captured CW to The Under Secretary of State and The War Office – reporting a total of 109,186 tons as of 14 August 1945. Including the 4,000 tons of bulk mustard/ arsinöl cited in the communication, the total was 113,186 tons.

SUMMARY OF CAPTURED CW - UK ZONE As of 14 August 1945				
	UK Depot #	Location	Corps District	CW Tonnage
1	18	Espelkampf	1	11,611
2	22	Rehden	30	3,632
3	24	Munsterlager	30	16,000
4	29	Celle	30	4,524
5	33	Orrel	30	33,600
6	35	Schneverden	30	723
7	38	Walsrode	30	5,600
8	50	Molln	8	3,935
9	74	Ilster	30	8,000
10	78	Bodenteich	30	5,279
11	79	Hannigsen	30	1,053
12	90	Sennelager	1	3,454
13	93	Diekholsen	30	3,775
14	109	Lehre	30	8,000
				109,186
	+ 4,000 Tons of Bulk Mustard/Arsinol			4,000
				113,186

Figure 51
Summary of Captured Chemical Weapons in the British Zone
As of 14 August 1945

Appendix 'A' to 21 AGp/1558/G(SDW 11) dated 23 Aug 45

PART 'A' - CW GERMAN AMMUNITION FILLED "Y"

	1 Corps District		8 Corps Dist	30 Corps District											
	Dump 18 W.6020	Dump 90 B.6853	Dump 50 S.9606	Dump 22 W.5047	Dump 24 X.5890	Dump 29 X.5949	Dump 33 X.6488	Dump 35 S.3405	Dump 38 X.2571	Dump 74 X.5393	Dump 78 X.9876	Dump 79 X.6135	Dump 93 L.5092	Dump 109 Y.0318	Totals
Shells, Mortars, Rockets Mines	8730	1892	2185	150	12300				2850		1904	73	3000	1550	34634
GAF Bombs					600						325				925
PART 'B' - CW GERMAN AMMUNITION FILLED "TABUN"															
Grun-ring 3	353										1066			2300	3719
PART 'C' - CW GERMAN AMMUNITION FILLING AS YET UNCHECKED															
Shells, Mortars, etc.					1300	4524		723		8000	1267				1581[illegible]
Bombs, 250 Kgs							33600								33600
PART 'D' - CW GERMAN AMMUNITION FILLING WHITE[illegible] AND BLAURING ([illegible] or DA)															
Shells, Mortars	2528	1562	1750	3482	1800				2750		717	980	775	4150	20494
Totals	11611	3454	3935	3632	16000	4524	33600	723	5600	8000	5279	1053	3775	8000	109186

Figure 52
Inventory of CW Captured in the British Zone
As of 25 August 1945

On 30 September 1945, the Control Commission for Germany (British Element) prepared their 15th Progress Report of covering the period 1-30 September 1945. The following information concerning CW disposal was excerpted from Appendix B, Army Division Progress Report:

"Army Division: 27,000 tons of CW ammunition has been dumped at sea out of a total so far uncovered of 120,000 tons."[73]

British Army of the Rhine (BAOR) Disarmament Reports provided the total of CW munitions and bulk stocks disposed of by the British authorities. The last report to mention CW stocks is that of December 1947 which states a total of 119,910 tons of chemical weapons had been disposed of from British control.[74]

BOTTOM LINE: It is concluded that 119,910 tons of CW were found in the British Zone. This number is the latest and most precise official number and is reflected in the Total Tonnage Table (Figure 55).

2.3.3 The Russian Zone

According to the October 1993 National Report of the Russian Federation entitled "Complex Analysis of the Hazard Related to the Captured German Chemical Weapon (sic) Dumped in the Baltic Sea", a total of 70,500 tons of chemical weapons were found in the Russian Zone.

Amounts of captured German CW found in zones of occupation (thousands of tons, for ammunition - including mass of metal):

Type of CW	Zone of Occupation			
	USSR	USA	Gr.Brit.	France
Chemical Ammunition	62.5	94.0	122.5	9.3
War Gases in encasements	8.0	10.5	4.2	0.2
Total	70.5	104.5	126.7	9.5
Σ		311.2		

Figure 53
Total CW Tonnage Captured in the Russian Zone[75]

The report also itemizes the exactly 608,482 individual pieces of captured German CW comprising the 70,500 tons – by munition type and contents. This data is presented in Figure 101, Chapter 3, Accounting of All Sea-Disposed Chemical Weapons.

What the October 1993 National Report of the Russian Federation does not divulge, however, are the Baltic Sea disposals of approximately 398,895 tons of excess Russian CW stocks which are suspected to have occurred until 1990. This number will also be addressed in Chapter 3.

BOTTOM LINE: A total of 70,500 tons of chemical weapons were found in the Russian Zone. This number is reflected in the Total Tonnage Table (Figure 55).

2.3.4 The French Zone

The following is the text of a declassified SECRET letter sent from HQ, British Army of the Rhine (BAOR) to The War Office on 27 February 1946. The subject was "CW Ammunition Held by the French":

"1. The following information concerning stocks of German CW ammunition held by the French has been communicated to this Headquarters by Lt. Col. D'Anselme, Director Generale du Controle da Desarmament, Commandement en Chef Francais en Allemagne, Baden-Baden.

2. A total quantity of 8,000 tons of CW ammunition is held by the French at the Urlau Depot (near to Constance), this being made up as follows:
(a) 2,000 tons, shell charged "DM",
(b) 3,000 tons, shell charged "H", and
(c) 3,000 tons, shell charged TABUN.

3. Items in para 2(a) and (b) are being destroyed by the French, while assistance is being sought from this HQ (HQ, BAOR) with regard to the destruction of the TABUN (sub-para 2(c)) by deep sea dumping.[76]

BOTTOM LINE: A total of 8,000 tons of chemical weapons were found in the French Zone. This number is reflected in the Total Tonnage Table. (Figure 55)

CHAPTER 3

ACCOUNTING OF ALL SEA-DISPOSED CHEMICAL WEAPONS

"The generals will never tell the truth: where and what weapons did they bury, develop, and dump. For this reason, I appeal to those who can divert a chemical Chernobyl from our home and to those who remember and understand, I implore you to respond! Before it is too late, it is necessary to draw up a chart of where chemical death lies in wait of its hour "X".

Russian Dr. of Chemical Sciences Lev Fedorov
1993

3.1 Estimated Total Chemical Weapons Disposed in European Waters Post-WWII

Chapter 3 is the nexus of the first research methodology protocol identified in the Introduction and summarized by the question: What chemical weapons were sea-disposed and where? It corroborates the total amount of CW disposed in European waters between the first British sea disposal operation in October 1945 and the final Russian Baltic disposals of excess CW stocks alleged to have occurred in 1989-90.

Figure 54 depicts the primary European bodies of water in which CW dumping is known to have occurred, to include the Baltic Sea, the Skagerrak and Kattegat, the North Sea, the Norwegian Sea, the Atlantic Ocean, the Bay of Biscay, and the Mediterranean Sea.

Figure 54
Map of the European Waters
in which CW Disposals Occurred

The sea disposal of World War II stocks of chemical weapons actually began well before the war officially ended. Aware that they were fighting a losing battle, the German command decided to try to move (and in some cases destroy) part of the CW stockpile, in particular tabun (GA), which Germany believed was still a secret weapon. Much of the stock was moved from areas about to be captured by the Allies as they advanced. Several secret transports took place, and some of the tabun (69,000 shells with an additional 5,000 tons of phosgene (CG) and tabun munitions) was dumped in the southern entrance to the Little Belt, Denmark in an operation during the last days of the war.[77]

These disposal operations are corroborated in the January 1994 Report to the 15h Meeting of the Helsinki Commission from the Ad-Hoc Working Group on Dumped Chemical Munition (sic):

> *Shortly before the end of WWII, German vessels loaded with nerve agent projectiles (tabun) and other types of chemical and conventional munitions were sunk at the southern entrance to Little Belt, approximately seven nautical miles south-east of Pøls Huk at the position 54°48'22"N, 10°13'22"E. Chemical warfare agents were also dumped in the sea south of Little Belt at the position 54° 47'-50'N, 10° 08-15'E.*[78]

According to the German Federal Maritime and Hydrographic Agency's May 1993 report on "Chemical Munitions in the Southern and Western Baltic Sea", in 1959 and 1960, "the tabun shells were recovered from the ships, set in concrete blocks, and re-dumped in the Bay of Biscay. Around 5,000 tons of bombs and shells containing tabun and phosgene were thus left in the area."[79] See Section 3.2.4 for additional detail on the German disposal operations.

Figures 55 and 56 provide upper and lower range summaries of the total CW tonnage sea disposed in European waters, by nation, respectively. The variable is the suspected Russian disposal of 112,523 tons of excess CW stocks in the Baltic during the period 1989-1990. This is addressed in paragraph 3.2.3, Russia – Disposal Operations.

TOTAL CW TONNAGE SEA DISPOSED IN EUROPEAN WATERS - BY NATION	TOTAL TONNAGE FOUND IN ZONE		TOTAL TONNAGE SEA DISPOSED	
GERMANY				
Disposed in the Baltic's Little Belt before WWII End			5,000	0.7%
69,000 Tabun (GA) Shells Re-Disposed in Bay of Biscay (1960)			Unk	
FRANCE	8,000			
- Destroyed in Place		-5,000		
- Transferred to UK for Sea Disposal		-3,000		
			0	0.0%
US	122,409			
- Destroyed in Place		-36,434		
- Shipped to US		-13,664		
- Shipped to Italy		-11,673		
- Transferred to UK for Sea Disposal		-10,436		
- Sold to German Industry		-9,464		
- Shipped to UK		-8,857		
- Shipped to Belgium		-2		
Skagerrak and North Sea Disposal Operations (1946-1948)			31,880	4.2%
Disposal Operations Off St. Raphael, France (1946)			1,700	
Disposal Operations Off the East and West Coasts of Italy (1945-1946)			Unk	
UK	119,910			
Skagerrak Disposal Operations (1945-1947)		127,000		
Atlantic Disposal Operations (1945-1956)		120,000		
			247,000	32.7%
RUSSIA	70,500			
Captured German CW and Excess Stocks (1946-1978)		356,872		
+ Suspected Excess Stocks (1989-1990)		112,523		
			469,395	62.2%
	320,819		**754,975**	
SUMMARY:	Short Tons (2,000 pounds/ton) (US): 754,975			
	Pounds: 1,509,950,000			
	Kilograms: 684,901,769			

Figure 55
Total Tonnage Sea Disposed, by Nation
Upper Range

Assuming the Russian disposals did occur in 1989-1990, the total CW tonnage sea-disposed in European waters between 1945 and 1990 was at least 754,975 tons, gross weight. This equates to over 1,509 million pounds and over 684 million kilograms.

TOTAL CW TONNAGE SEA DISPOSED IN EUROPEAN WATERS - BY NATION		TOTAL TONNAGE FOUND IN ZONE		TOTAL TONNAGE SEA DISPOSED	
GERMANY					
	Disposed in the Baltic's Little Belt before WWII End			5,000	0.8%
	69,000 Tabun (GA) Shells Re-Disposed in Bay of Biscay (1960)			Unk	
FRANCE		8,000			
	- Destroyed in Place		-5,000		
	- Transferred to UK for Sea Disposal		-3,000		
				0	0.0%
US		122,409			
	- Destroyed in Place		-36,434		
	- Shipped to US		-13,664		
	- Shipped to Italy		-11,673		
	- Transferred to UK for Sea Disposal		-10,436		
	- Sold to German Industry		-9,464		
	- Shipped to UK		-8,857		
	- Shipped to Belgium		-2		
	Skagerrak and North Sea Disposal Operations (1946-1948)			31,880	5.0%
	Disposal Operations Off St. Raphael, France (1946)			1,700	
	Disposal Operations Off the East and West Coasts of Italy (1945-1946)			Unk	
UK		119,910			
	Skagerrak Disposal Operations (1945-1947)		127,000		
	Atlantic Disposal Operations (1945-1956)		120,000		
				247,000	38.4%
RUSSIA		70,500			
	Captured German CW and Excess Stocks (1946-1978)		356,872		
	+ Suspected Excess Stocks (1989-1990)		0		
				356,872	55.5%
		320,819		**642,452**	
SUMMARY:	Short Tons (2,000 pounds/ton) (US):	642,452			
	Pounds:	1,284,904,000			
	Kilograms:	582,822,625			

Figure 56
Total Tonnage Sea Disposed, by Nation
Lower Range

Assuming the Russian disposals did NOT occur in 1989-1990, the total CW tonnage sea-disposed in European waters between 1945 and 1990 was at least 642,452 tons, gross weight. This equates to over 1,284 million pounds and over 582 million kilograms.

3.2 Disposal Operations

To account for the total tonnage of chemical weapons sea-disposed in European waters post World War II, each of the principal dumping States will be examined separately – the United States (the disposal of captured German CW unwanted by the US and UK in Operation "Davey Jones Locker" conducted in the Skagerrak and Norwegian Sea and the disposal of excess U.S. CW stocks off of Italy), the United Kingdom (with both Skagerrak and Atlantic operations), Russia (with loose dumping operations in the Baltic), and Germany (with operations conducted immediately before the war ended).

3.2.1 United States – Disposal Operations

Captured German CW – Operation "Davey Jones Locker"

Between July 1, 1946 and August 24, 1948, in an operation code-named "Davey Jones Locker", the United Stated scuttled eleven ships containing a total of 31,880 tons of chemical weapons – 26.0% of the total tonnage found in the American Zone. CW cargoes ranged from 751-6,720 tons. The average weight was 2,898 tons. Figure 57 lists each of the hulks scuttled by the United States in the Skagerrak and Norwegian Sea by date and includes CW tonnage, depth, and location.

The primary sources of information and data regarding Operation "Davey Jones Locker" include the following:

- "The History of Captured Enemy Toxic Munitions in the American Zone, European Theater, May 1945 to June 1947" published by the Office of the Chief of Chemical Corps, U.S. Army European Command in 1947.

- "Addendum to the History of Captured Enemy Toxic Munitions in the American Zone, European Command, June 1947 to August 1948" published by the Office of the Chief, Chemical Division, Headquarters European Command in January 1949.
- "Meeting Notes: Summary of Some Chemical Munitions Sea Dumps by the United States" and "Inventory of Toxic Chemical and Biological (Toxin) Agents, Munitions, Dumps, and Historical Summary" – two official declassified documents from the U.S. Army Soldier and Biological Chemical Command (SBCCOM), Edgewood Arsenal, Maryland.

Figures 58 and 59 are maps displaying the reported locations of the nine ships scuttled in the Skagerrak (U.S. Convoys DJL 1-DJL 7).

SUMMARY OF CHEMICAL WEAPONS SCUTTLED IN THE SKAGERRAK AND NORTH SEA BY THE UNITED STATES IN "OPERATION DAVEY JONES LOCKER"

1 July 1946 - 24 August 1948

	CONVOY	Ship NAME	Ship TYPE	SHORT TONS CM LOADED	DATE SCUTTLED	DEPTH (Meters)	LATITUDE	LONGITUDE
1	DJL1	SPERRBRECHER	German Minebreaker	1,510.9	01-Jul-46	650	58 14'00"	09 15'00"
2	DJL1	T-65	German Flak Ship	1,709.1	01-Jul-46	650	58 17'09"	09 37'01"
3	DJL1	U-J 305	German Trawler	751.5	02-Jul-46	650	58 16'04"	09 29'00"
4	DJL2	ALCO BANNER	Hog-Islander	3,096.8	14-Jul-46	650	58 18'07"	09 36'05"
5	DJL3	JAMES OTIS	American Liberty	4,091.4	30-Aug-46	680	58 16'00"	09 32'00"
6	DJL4	JAMES SEWELL	American Liberty	4,480.0	06-Jun-47	725	58 15'02"	09 30'06"
7	DJL5	JAMES HARROD	American Liberty	3,360.0	20-Jun-47	665	58 16'00"	09 33'00"
8	DJL6	GEORGE HAWLEY	American Liberty	1,120.0	30-Jun-47	685	58 18'05"	09 38'00"
9	DJL7	NESBITT	American Liberty	6,720.0	18-Jul-47	580	58 15'00"	09 30'00"
10	DJL8	PHILIP HEINIKEN	German Freighter	2,240.0	24-Jul-48	1035	62 57'00"	01 32'00"
11	DJL9	MARCY	German Freighter	2,800.0	24-Aug-48	1180	62 59'00"	01 23'00"
			TOTAL:	31,879.7				

	TONS	DATES	DAYS
DJL 1-3 (1946)	11,159.7	1 Jul - 30 Aug 46	60
DJL 4-7 (1947)	15,680.0	6 Jun - 18 Jul 47	42
DJL 8-9 (1948)	5,040.0	24 Jul - 24 Aug 48	31
	31,879.7		

DJL 8-9 (1948) 15.8%

DJL 1-3 (1946) 35.0%

DJL 4-7 (1947) 49.2%

Figure 57
Details of the 11 Ships Scuttled in
Operation "Davey Jones Locker"
July 1, 1946 - August 24, 1948

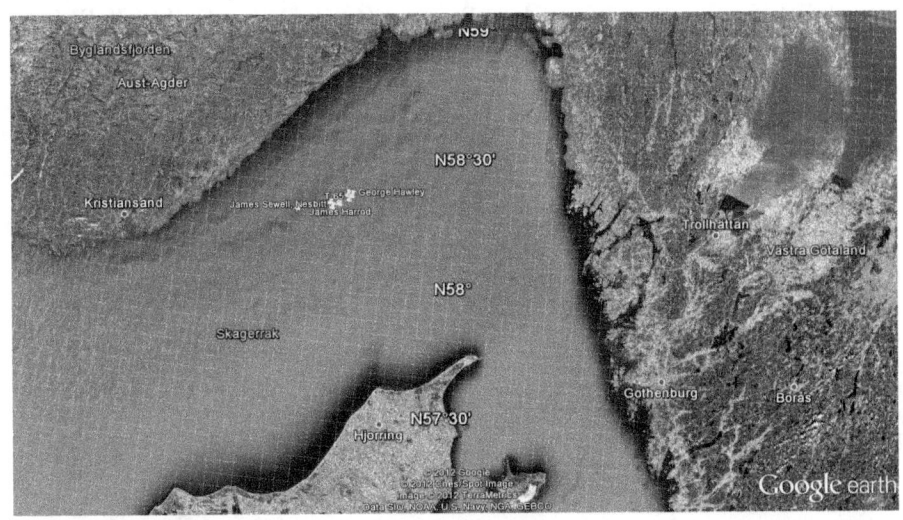

Figure 58
Map 1 of the Nine CW-Laden Ships Scuttled in the Skagerrak,
1946-47, in Operation "Davey Jones Locker"

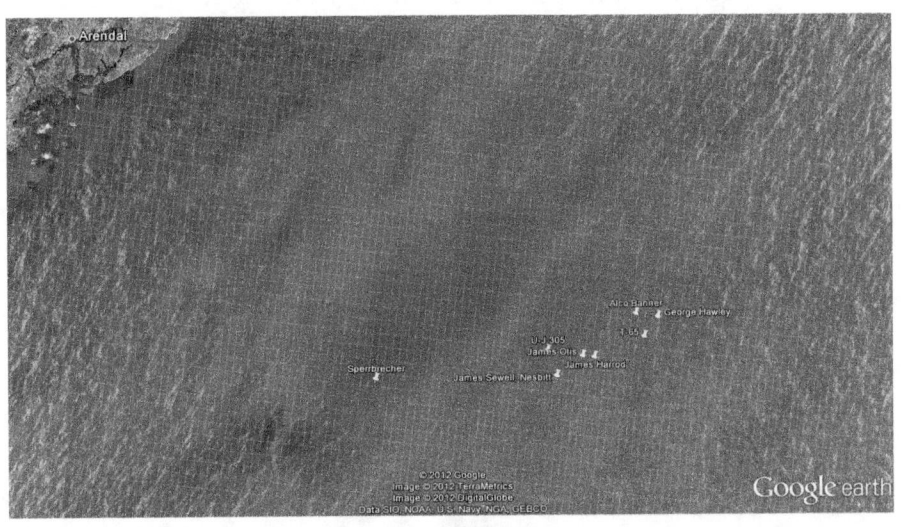

Figure 59
Map 2 of the Nine CW-Laden Ships Scuttled in the Skagerrak,
1946-47, in Operation "Davey Jones Locker"

Figures 60 and 61 are maps displaying the reported locations of the two ships scuttled in the Norwegian Sea in 1948 (U.S. Convoys DJL 8 and DJL 9).

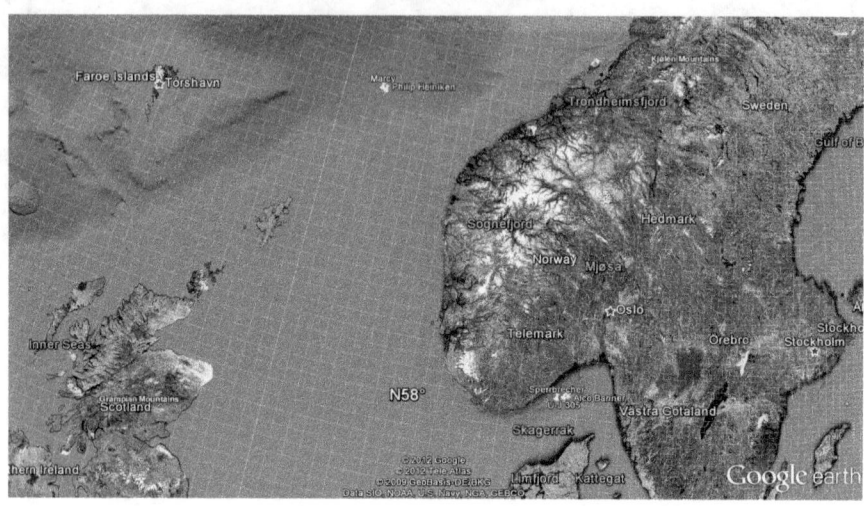

Figure 60
Map 1 of the Two CW-Laden Ships Scuttled in the Norwegian Sea, 1948, in Operation "Davey Jones Locker"

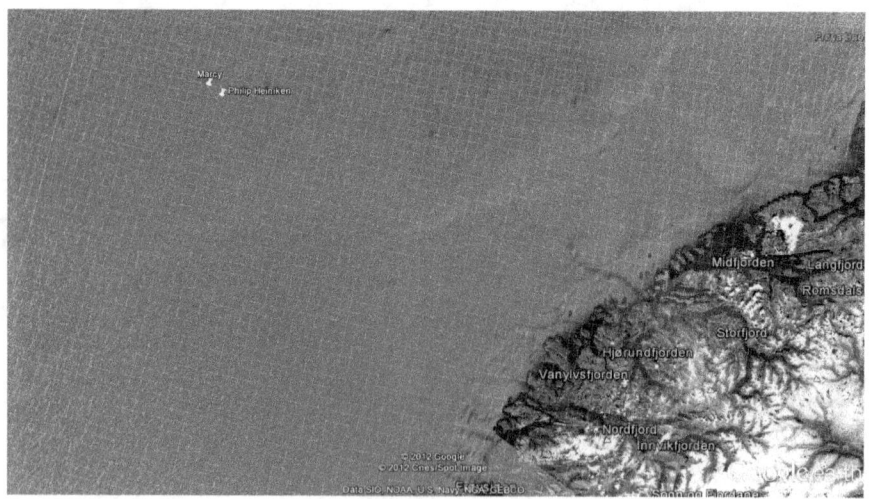

Figure 61
Map 2 of the Two CW-Laden Ships Scuttled in the Norwegian Sea, 1948, in Operation "Davey Jones Locker"

By June 1946, all U.S. requirements for CW from captured German stocks had been shipped, destruction of chemical weapons at depots in situ was underway, and initial convoys of loaded hulks had started.[80]

Figure 62
Operation "Davey Jones Locker"
Unloading Rail Wagons, 12 June 1946

A loading operation showing 15cm Tabun-filled poison gas shells being transferred from the freight cars to pallets for transfer to ships at Midgard Docks. Chemical Warfare Service officers are inspecting the operation taking place.

Figure 63
Operation "Davey Jones Locker" – Inspection of Wagons
Loaded with Box Type Crates, 12 June 1946

A U.S. Army soldier is supervising German civilian laborers in the loading of German poison gas shells from the freight car to the hold of the hulk T-65 at the Midgard docks of the Nordenham Sub Port of the 17th Major Port Command.

Figure 64
Operation "Davey Jones Locker"
Loaded Hulks Ready for Sea, 2 July 1946

A view of the Midgard Docks showing railroad siding, a mobile crane, and the loaded hulks T-65 and the Sperrbrecher.

The first American chemical weapons convoy of "Operation Davey Jones Locker" occurred on 1-2 July 1946. Note the first British chemical weapons sea disposal operation in the Atlantic occurred a year earlier on 2 July 1945 with the scuttling of the Empire Fal. The first British CW sea disposal operation in the Skagerrak (CW1) occurred 9 months earlier, on October 4, 1945.

The first two ships of the three-ship U.S. convoy, the Sperrbrecher and the T-65, were scuttled in the Skagerrak with a combined total of 3,220 tons of munitions containing tabun (GA), mustard (H), phosgene (CG), and chloroacetophenone (CNB).

Figure 65
Operation "Davey Jones Locker"
First Convoy at Sea, 1-2 July 1946

At 3:55pm on 1 July 1946, the Sperrbrecher, a German minebreaker loaded with 1,511 tons of toxic munitions, was scuttled at 58°14'N, 09°15'E at a depth of 650 meters. Sinking was accomplished by opening of seacocks and by shellfire from the escort vessel. The Sperrbrecher was towed to sea by the German tug Sirus and an American tug LT-159, the John O'Reilly.

Figure 66
Operation "Davey Jones Locker"
The Sperrbrecher, 1 July 1946

Later that day, at 8:25pm, the T-65, a former German flak ship loaded with 1,709 tons of toxic munitions was scuttled at 58°17'9"N, 09°37'1"E – also at a depth of 650 meters. The T-65 was towed to sea by the American tug LT-119. Sinking was accomplished by opening of seacocks and by shellfire from the escort vessel. Shells were fired into the hull at the water line in the bow and stern where no toxics were stowed. The escort vessel was upwind from the hulk.

Figure 67
Operation "Davey Jones Locker"
The T-65, 1 July 1946

The next day, at 12:37am, the U.J.-305, a German trawler loaded with 752 tons of CW, was scuttled at 58°16'4"N, 09°29'E. The U.J.-305 was towed to sea by the German tug Widder. The seacocks were opened and shells were fired into the ship's hull at the water line in the bow and stern where no toxics were stored. According to official records, the U.J.-305 is at 650 meters.

Figure 68
Operation "Davey Jones Locker"
The U-J 305, 2 July 1946

The first scuttling of the second Operation "Davey Jones Locker" convoy occurred at 9:00am on 14 July 1946. The Alco Banner, a hog-islander loaded with 3,097 tons of chemical weapons, was sunk in the Skagerrak at 58°18'7"N, 09°36'5"E. 105mm armor-piercing shells were fired from a German torpedo boat, from which the picture below was taken during the firing of a round. It took 45 minutes for the Alco Banner to sink. According to official records, the Alco Banner is at 650 meters.

Figure 69
Operation "Davey Jones Locker"
The Alco Banner, 14 July 1946

In the third convoy of Operation "Davey Jones Locker", the 5th of 11 ships scuttled, the James Otis – an American liberty ship loaded with 4,091 tons of chemical weapons – was sunk in the Skagerrak at 10:17am on 30 August 1946 (58°16'N, 09°32'E) at a depth of 650 meters.[81] This was the last U.S. chemical weapons convoy in 1946. It is not known how the James Otis was sent to the bottom.

Figure 70
Operation "Davey Jones Locker"
The James Otis, 30 August 1946

In the first U.S. sea disposal operation of 1947, the James Sewell, an American liberty ship loaded with 4,480 tons of chemical weapons, was scuttled at 12:23pm, 6 June in the Skagerrak at 58°15'02"N, 09°30'06"E at a depth of 725 meters. Dumping had resumed after a halt of 280 days (over 9 months). The previous hulk was scuttled on 30 Aug 46.

Two weeks later, at 11:07am, 20 June 1947, the 7th of 11 vessels in Operation "Davey Jones Locker" was scuttled. The James Harrod, an American liberty ship loaded with 3,360 tons of chemical weapons, was sunk in the Skagerrak at 58°16'N, 09°33'E at a depth of 665 meters.[82]

Just 10 days after the 7th vessel was sunk, the George Hawley, an American liberty ship loaded with 1,120 tons of chemical weapons was sunk at 58°18'05"N, 09°38'E. The scuttling occurred at 1:25pm, 30 June 1947 at a depth of 685 meters.[83]

The final U.S. CW disposal in the Skagerrak took place at 9:24pm on 18 July 1947. The Nesbitt, an American liberty ship loaded with 6,720 tons of chemical weapons, was scuttled at 58°15'N 09°30'E at a depth of 580 meters.[84] This was the last U.S. disposal operation conducted in 1947. As the Nesbitt was being towed to sea, a heavy wind and tide combined to beach her on a reef at the mouth of the Weser River. The picture below shows the hulk being eased out to sea once again by a few of the many tugs that answered the emergency call.[85] The remaining two ships of Operation "Davey Jones Locker" were scuttled in the Norwegian Sea over a year later.

Figure 71
Operation "Davey Jones Locker"
The Nesbitt Being Towed to Sea, July 1947

Figure 72
Operation "Davey Jones Locker"
The Nesbitt Being Scuttled, 18 July 1947

Over a year after the final U.S. scuttling in the Skagerrak, vessels 10 and 11 in Operation "Davey Jones Locker" were sunk in the Norwegian Sea. At 8:28pm on 24 July 1948, the Philip Heiniken, a German freighter loaded with 2,240 tons of chemical weapons, was scuttled at 62°57'N, 01°32'E at a depth of 1,035 meters.[86] The picture below shows the Philip Heiniken being towed to sea by a German tug. Tugs were German-manned and under the command of U.S. Naval officers. The trip to the designated scuttling area required three to four days.

Figure 73
Operation "Davey Jones Locker"
The Phillip Heiniken, 24 July 1948

The 11th and final vessel to be scuttled in Operation "Davey Jones Locker" was the Marcy. At 5:15pm on 24 August 1948 the German freighter, loaded with 2,800 tons of chemical weapons was sunk at 62°59'N, 01°23'E at a depth of 1,180 meters.[87]

Figure 74 shows German laborers stacking 250 kg aerial bombs in hatch #1 of the ship. Workers were furnished impermeable and impregnated suits, gloves, overshoes, and gas masks by the Port Chemical Officer. Gas masks were maintained nearby, but were used only when required by the toxic concentration. All personnel involved were instructed and drilled in proper handling and safety measures by the supervising Chemical Corps specialists.[88]

Figure 75 shows hatch #2 of the Marcy partially loaded with mustard-filled 55-gallon drums, shells, and rockets.[89]

Figure 76 shows the loaded hulk Marcy being towed out the Weser River to be scuttled in the Norwegian Sea. White circles were painted near the water line as targets for naval gunners. Fire and disaster plans were prepared for the toxic out-loading project at the Midgard Docks in Nordenham, Germany. All military and civilian personnel were adequately familiarized with the program and all emergency plans. Precautions included the stand-by of firefighting equipment, specially trained medical teams on duty at local hospitals, German police to keep all unauthorized persons away from the dock area and evacuation drills of the civilian population.[90]

11 ships, with a combined total of 31,880 tons of chemical warfare ammunition, were scuttled by the United States in this 785-day operation.

Figures 74 and 75
Operation "Davey Jones Locker"
Loading the Marcy and Hatch #2

Figure 76
Operation "Davey Jones Locker"
The Marcy

MEDITERRANEAN DISPOSAL OPERATIONS

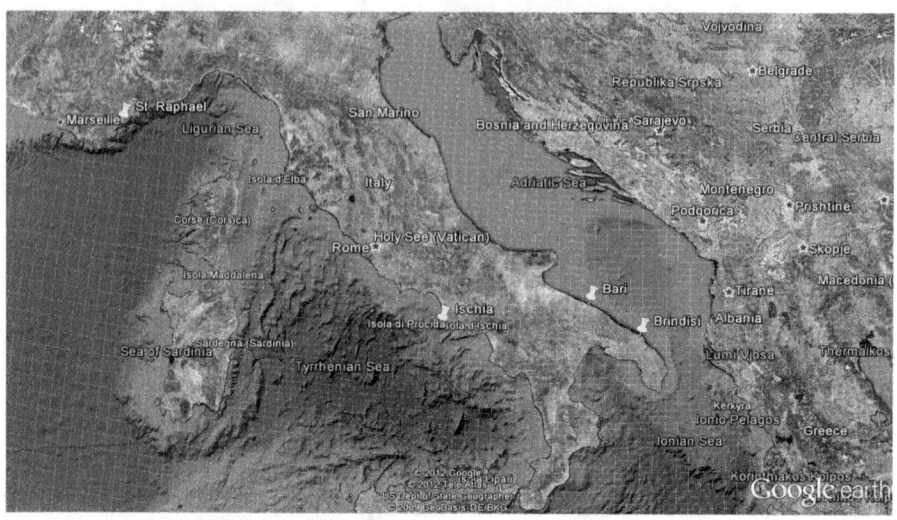

Figure 77
Map of U.S. Disposal Locations in the Mediterranean Sea

Disposal Operations Off of Italy

According to official U.S. records, unspecified amounts of Hydrocyanic Acid / Hydrogen Cyanide (AC) (blood agent), Phosgene (CG) (lung poison/choking agent), Cyanogen Chloride (CK) (blood agent), Mustard (H) (blistering agent), and Lewisite (L) (blistering agent) bombs which had been stored in U.S. Army depots in Italy, were loose dumped off the west coast of Italy (off the island of Ischia in the Bay of Naples) and the east coast of Italy – near Bari and Brindisi.[91]

It is evident from existing U.S. Army records there was either confusion regarding locations or inaccurate record-keeping. Exemplary of this is the document at Figure 78. The island of Ischia is on the west coast of Italy; Bari is on the east coast.

```
16.  FROM:      Italian Depots (U.S. Army)
     TO:        Mediterranean Sea off Italy (Island of Ischia, near Bari)
     DATE:      21 October – 5 November 1945, 1-15 December 1945
     ACTION:    Ocean Disposal
     MATERIAL:  Bombs (CG), unspecified quantity
                Bombs (CK), unspecified quantity
                Bombs (AC), unspecified quantity




22.  FROM:      Auera, Italy
     TO:        Mediterranean Sea off Italy (Island of Ischia, near Bari)
     DATE:      1-23 April 1946
     ACTION:    Ocean Disposal
     MATERIAL:  Bombs (L and/or H), unspecified quantity
```

Figure 78
U.S. Army Record of Disposal Operations
Off the Coast of Ischia (Naples) and Bari, Italy[92]

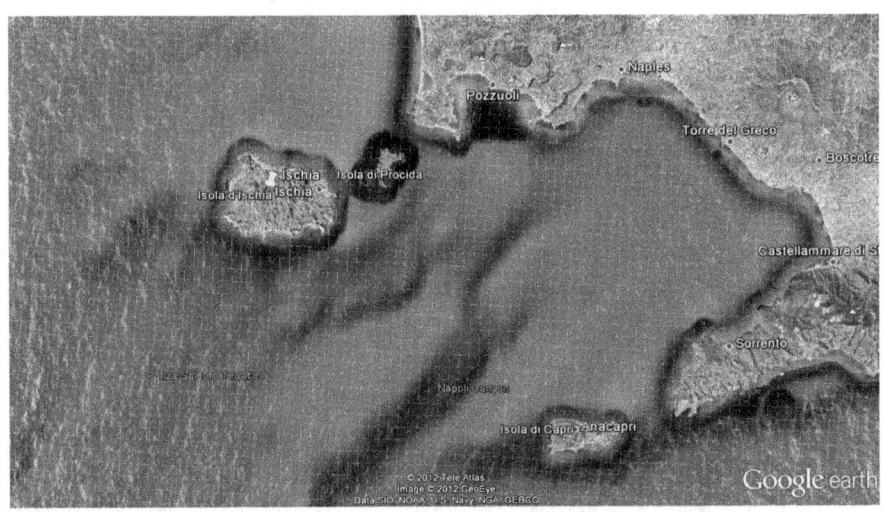

Figure 79
Map of U.S. Disposal Locations
Off the West Coast of Italy

Specifically regarding disposal operations in the Bay of Naples, the following text is excerpted from the U.S. Army Center of Military History's historical volume "The Chemical Warfare Service: Chemicals in Combat", first published in 1966:

"In Italy, the 92nd Chemical Composite Company served primarily as a depot unit. Upon its arrival in mid-August 1944, it was sent to the Chemical Warfare Service depot near Bagnoli. For the remainder of the war, the 92nd remained on the job as a depot-maintenance outfit at Bagnoli and Leghorn.

The cessation of hostilities in Europe did not reduce the company's workload for some time. The Leghorn detachment, much depleted by personnel transfers, acquired control of a German prisoner of war battalion to help it rehabilitate, box, and ship materiel to the Far East, a task accomplished ahead of schedule.

The Bagnoli portion of the company celebrated the end of the war by disposing of its supply of toxics through the winter of 1945-46, sinking the materiel in deep water off the island of Ischia.

It was not until the spring of 1946 that the 92nd, by then possessed of a Meritorious Service Unit Plaque, was ready for inactivation."[93]

The Chemical Disaster at Bari – "The Second Pearl Harbor"

Also excerpted from the U.S. Army Center of Military History's historical volume "The Chemical Warfare Service: Chemicals in Combat":

"If gas warfare had ever been a threat in the Mediterranean area, it would have been at the time of the assault landings. In the Sicilian landings small stores of enemy toxics were found, but their placement and manner of storage indicated that there was no intention of using them. For the landings on the Italian mainland, Allied intelligence officers feared that toxics would be employed by the enemy, and as a result, retaliatory stocks were brought in too soon. A tragic gassing of Allied forces in the harbor of Bari, Italy occurred when enemy action breached a ship carrying Allied gas."[94] Here are the details:

Sources of information concerning the German Luftwaffe attack on the Italian port of Bari on December 2, 1943 include U.S. Arms Control and Disarmament Agency (ACDA), U.S. Department of State's 1993 "Special Study on the Sea Disposal of Chemical Munitions", The History Net's research on "Bari: The Second Pearl Harbor" and Gerald Reminick's book "Nightmare in Bari: The World War II Liberty Ship Poison Gas Disaster and Cover-Up".

"With the rest of the world, President Franklin D. Roosevelt opposed chemical warfare, but knew the necessity of being prepared should the enemy use it.

Retaliation meant having the weapons on hand in various theaters of battle. Chemical gas supplies were therefore necessary in the event Germany decided to initiate it use against the Allied thrust up the Italian peninsula and into Europe. Bari, now under Allied control, was the logical port of entry for the deadly chemicals. The port was under the jurisdiction of the British, in part because Bari was the main supply base for General Bernard Law Montgomery's Eighth Army. The city was also the newly designated headquarters for the American Fifteenth Air Force which had been activated in November 1943.

To ensure potential retaliatory capability, President Roosevelt authorized the confidential shipment of 2,000 (M47A2) 100 pound mustard gas bombs to Bari. The bombs, measuring four feet long by 18 inches in diameter, were manufactured at Edgewood Arsenal, Maryland. They were shipped through the eastern Chemical Warfare Depot to Curtis Bay Depot of Baltimore by train and into the port of Baltimore for loading aboard ship.

First Lieutenant Howard D. Beckstrom was placed in charge of six enlisted men from the 701st Chemical Maintenance Company. His job was to shepherd the bombs to their final destination. Lieutenant Thomas A. Richardson was assigned the role of cargo officer in charge of security.

The Liberty Ship SS John Harvey was selected to transport the chemical munitions to Bari, Italy. Commencing 30 September 1943, the mustard gas bombs were loaded and on 8 October she sailed to Norfolk, Virginia arriving on 9 October. On 15 October she sailed from Norfolk arriving in Oran, Algeria on 2 November. While in Oran, some cargo was discharged. The mustard gas bombs remained, however. Evidence indicates additional cargo consisting of ammunition was loaded.

On 20 November the John Harvey departed Oran in a convoy of 40 ships and arrived in Port Augusta, Sicily on 25 November. The next day the ship departed in another convoy of 30 ships with half of this number sailing to Taranto – a port in southern Italy which is now a main

Italian Naval Base. The John Harvey arrived in Bari, Italy on Sunday, 28 November and was moored at Berth 29 on the east jetty of the harbor.

Over the next five days, cargo security officer Lieutenant Richardson made a number of attempts to have the CW cargo unloaded. Captain Elwin Foster Knowles, the commanding officer of the John Harvey, even accompanied Richardson on several occasions to encourage the authorities at the War Shipping Administration Office to expedite the removal of the cargo, but to no avail. Since the lethal gas was not officially on board, the John Harvey was not about to be given special priority.

On Thursday, 2 December 1943, approximately 50 ships lay in the busy port.

Totally absorbed by the task of getting the Fifteenth Air Force off the ground, the Allies gave little thought to the possibility of a German air raid on Bari. The Luftwaffe in Italy was relatively weak and stretched so thin it could hardly mount a major effort. Or so Allied leaders believed.

German reconnaissance flights over Bari were seen as a nuisance. At first, British anti-aircraft batteries fired a half-hearted round or two, but eventually they ignored the German flights altogether. Why waste ammunition?

Responding to rumblings about lax security measures, British Air Vice Marshal Sir Arthur Coningham held a press conference on the afternoon of 2 December 1943 and assured reporters that the Luftwaffe was defeated in Italy. He was confident the Germans would never attack Bari.

Not everyone was as sure the German air force was so crippled, however. British Army Captain A.B. Jenks, who was responsible for the port's defense, knew that preparations for an attack were woefully inadequate. But his voice, as well as those of one or two others, was drowned out by a chorus of complacent officers. When darkness came, Bari's docks were brilliantly lit so unloading of cargo could continue. Little thought was given to the need for a blackout.

From the German perspective, the Bari attack was the product of a planning session between Luftwaffe Field Marshal Albert Kesselring and his subordinates. The Allied airfields at Foggia were discussed as possible targets, but Luftwaffe resources were stretched too thin to permit the effective bombing of such a large complex of targets.

It was Field Marshal Wolfram von Richthofen, commander of Luftflotte 2, who suggested Bari as an alternative. His advice, Kesselring knew, was sound. Richthofen believed that if the port was crippled, the British Eights Army's advance might be slowed and the nascent American Fifteenth Air Force's bomber offensive delayed. Richthofen told Kesselring that the only planes available for such a task were his Junkers Ju-88 A-4 bombers. With luck, he might scrape together 150 such planes for the raid.

When the strike force was mustered, however, there were only 105 Ju-88s available for the mission. But the element of surprise, coupled with an attack at dusk, might shift the odds in the Germans' favor. Most of the planes would come from Italy, but Richthofen purposely wanted to obfuscate matters by using a few Ju-88s from

Yugoslavia. If the Allies thought the entire mission originated from there, they just might misdirect retaliatory strikes to the Balkans.

The Ju-88 pilots were ordered to fly their twin-engine bombers east to the Adriatic Sea, then swing south and west. British anti-aircraft would probably expect an attack, if any, to come from the north, not from the west. The Ju-88s were also supplied with Duppel – thin strips of tinfoil cut to various lengths. The tinfoil registered like aircraft on radar screens, producing scores of phantom targets.

The aim of the German pilots was to arrive over Bari around 7:30 p.m. Parachute flares would be released first to light the way for the attacking aircraft, and the Ju-88s would come in low, trying to get under Allied radar that was already confused by the Duppel.

The Germans arrived at Bari on schedule. First Lieutenant Gustav Teuber, leading the first wave, could hardly believe his eyes. The docks were brilliantly lit; cranes stood out in sharp relief as they unloaded cargo from the ships' gaping holds, and the east jetty was packed with ships – including the SS John Harvey.

The success of the Luftwaffe attack was stunning. In the span of only 20 minutes, the raid became the worst bombing of Allied shipping since Pearl Harbor two years earlier. A total of seventeen Allied ships were destroyed – five U.S. Liberty, four British, three Norwegian, three Italian, and two Polish. Six other merchant ships were seriously damaged.

The SS John Harvey (one of the five U.S. Liberty ships in the harbor at the time) with its Top Secret cargo of approximately 100 tons of mustard gas bombs, received a direct hit, exploded, and disappeared in a huge, mushroom-shaped fireball that hurled pieces of the ship and her cargo hundreds of feet into the air. Everyone on board was killed instantly. The blast sent out multihued fingers of smoke and sent the toxic chemical across the water and vaporizing it through the air of Bari.

The blast also completely ripped apart the mass of inhabitants of the Old City that had crowded along the shore. Behind them, houses were folded up like cardboard doll-houses and collapsed into heaps of rubble. Many died without knowing what happened. Farther away, the blast knocked down people in streets, unhinged doors and shutters, and blew in the windows of almost every building in the city. The concussion was felt at some points 20 miles or more from the harbor.

The loss of life was appalling. More than 1,000 Allied servicemen and more than 1,000 civilians were killed. The total number of deaths will never be known. In the horror and confusion, records of treatment were poorly kept, if at all. In addition, many survivors of the initial bombing later died from toxic contamination – but their deaths were not attributed to the attack. To this day, few people know of the disaster at Bari.

After the attack, the Allied hospitals occupying the Bari Polyclinic – still not fully equipped because many of the supplies they needed were on the damaged or destroyed Liberty ships – began receiving overwhelming numbers of casualties. Hospital units at this facility included the 3rd New Zealand General, 98th British

General, 14th Combined General, 30th Indian General, 34th Field Hygiene Section and the 4th Base Depot Medical Stores. The American 26th General Hospital was nearby.

Many survivors who spent time in the hospital never knew they had been exposed to the chemical agent until many years later, if at all. Many, many more will have died since of related illnesses such as cancer, leukemia, and bronchial and chest and throat ailments.

Besides military casualties, the citizens of Bari also needed vital medical attention. Unfortunately, they were last in order for treatment. No records were kept of casualties or deaths involving the people of Bari. As cited earlier, it was estimated that more than 1,000 Bari citizens were killed in the attack. How many people went untreated for mustard gas inhalation and burns, or would eventually develop complications, will never be known.

Questions abounded. Why had the attack occurred and why were the wounded showing symptoms of mysterious burns and sickness? No answers were forthcoming. The British, in charge of the port, denied they knew anything about mustard gas. In fact, Prime Minister Winston Churchill continued denying the fact even years after the war ended."

While there is a possibility some of the mustard gas bombs aboard the John Harvey survived the explosion and remain in Bari harbor, official U.S. records indicate that an unspecified amount of mustard gas bombs were disposed in the Adriatic Sea, five miles off the shore of Brindisi, Italy. Brinidisi is southeast of Bari on the Salento Peninsula.

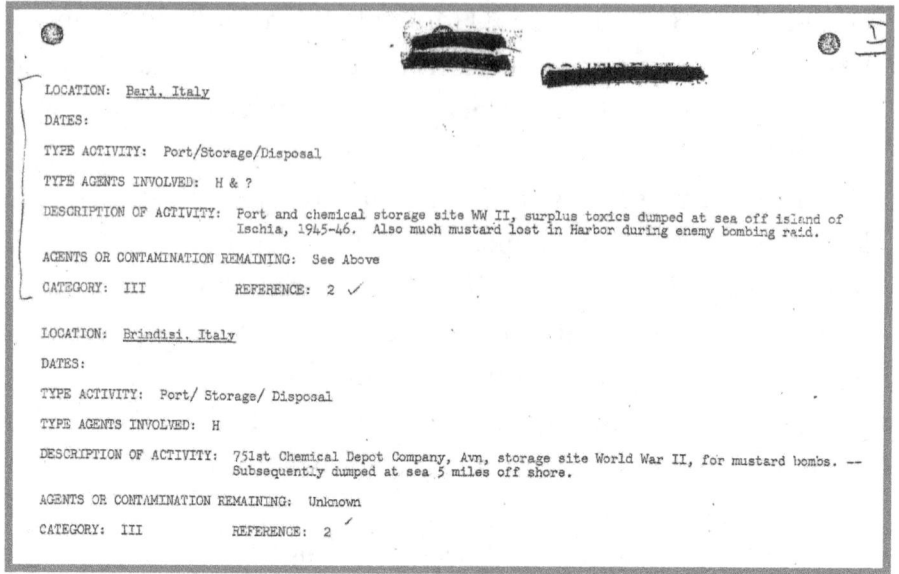

LOCATION: Bari, Italy

DATES:

TYPE ACTIVITY: Port/Storage/Disposal

TYPE AGENTS INVOLVED: H & ?

DESCRIPTION OF ACTIVITY: Port and chemical storage site WW II, surplus toxics dumped at sea off island of Ischia, 1945-46. Also much mustard lost in Harbor during enemy bombing raid.

AGENTS OR CONTAMINATION REMAINING: See Above

CATEGORY: III   REFERENCE: 2 ✓

LOCATION: Brindisi, Italy

DATES:

TYPE ACTIVITY: Port/ Storage/ Disposal

TYPE AGENTS INVOLVED: H

DESCRIPTION OF ACTIVITY: 751st Chemical Depot Company, Avn, storage site World War II, for mustard bombs. — Subsequently dumped at sea 5 miles off shore.

AGENTS OR CONTAMINATION REMAINING: Unknown

CATEGORY: III   REFERENCE: 2

Figure 80
Declassified SECRET Record of Disposal Operations Off the
Coast of Brindisi, Italy[95]

Disposal Operations off of St. Raphael, France

Official U.S. records also indicate that between July and October 1946, 1,700 tons of mustard and lewisite-filled bombs were disposed off the coast of St. Raphael, France, on the French Riviera.

St. Raphael website: **http://www.saint-raphael.com**

SEA DUMP OTHER THAN CHASE

DATE OF REPORT 25 AUG 1970
REPORTING AGENCY Edgewood Arsenal, Maryland

CONTAMINATED RANGES & BLDGS			KNOWLEDGE OF BURIAL SITES, BURNING PITS, CONTAM BLDGS, SEA DUMPS, OR OTHER DISPOSAL OR STORAGE AREAS		OTHER INFORMATION
			SOURCE		
RANGE	BLDG	AGENT	HEARSAY & PERSON	DOCUMENT & LOCATION	
				Surety Office Col. Dickey	From July - October 1946 there were 1700 tons of mustard and Lewisite filled bombs dumped off the Mediterranean shore at St. Raphael, France.
					Bari, Italy - This was a bombing of the harbor of Bari, Italy, where a ship with mustard filled bombs was hit at dockside and several deaths and casualties occurred.

Figure 81
Declassified SECRET Record of Disposal Operations
Off the Coast of St. Raphael, France[96]

Bottom Line: The United States disposed of 1,700 tons of mustard and lewisite-filled bombs off of St. Raphael, France between July and October 1946 and 31,880 tons of chemical weapons in the Skagerrak and Norwegian Sea between 1 July 1946 and 24 August 1948. Unspecified amounts of chemical weapons were also dumped in the Mediterranean Sea between 1945 and 1946 off the west coast of Italy near Naples and off the east coast near both Bari and Brindisi.

3.2.2 United Kingdom – Disposal Operations

According to Ministry of Defence (MoD) records, between 1945 and 1956 "at least" 247,000 tons of chemical weapons were disposed in the Skagerrak and Atlantic by the United Kingdom – 127,000 tons in the Skagerrak and 120,000 tons in the Atlantic.

As cited in the MoD's 1993 "Report of Sea Dumping of CW by the UK in the Skagerrak Waters Post WWII", "the surviving records of the sea-dumping programme indicate that <u>at least</u> 127,000 tons of CW munitions were disposed in the Skagerrak. This figure includes 3,000 tons from the French Zone and up to 10,000 tons from the American Zone, these latter quantities being sea-dumped by Britain on their Allies' behalf. 10,000 tons of CW munitions recovered in the British Zone were shipped to the U.K. for research purposes under the code name "Op Dismal".[97]

Research has shown, however, that the 247,000 ton total includes the majority of the 119,910 tons of CW captured in the British Zone of Occupation, 10,436 tons of CW captured in the American Zone and transferred to the UK for disposal, 3,000 tons of CW captured in the French Zone and transferred to the UK for disposal, 24,958 tons of excess CW stocks U.S. forces had in the European Theater which were secretly transferred to the UK for disposal, and excess UK chemical weapons stocks.

Although there are at least 20 additional vessels suspected to have been scuttled with chemical weapon cargoes during this period, 56 vessels scuttled by the UK are detailed in the sections below which address the Skagerrak and Atlantic disposal operations. Specific information on the locations of the scuttled ships is covered in Chapter 4.

Skagerrak Disposal Operations

Between 4 October 1945 and 6 June 1947, 32 vessels containing a total of at least 127,000 tons of chemical weapons were scuttled in the Skagerrak and Kattegat by the United Kingdom. Figures 82 and 83 below display the locations of the scuttled ships:

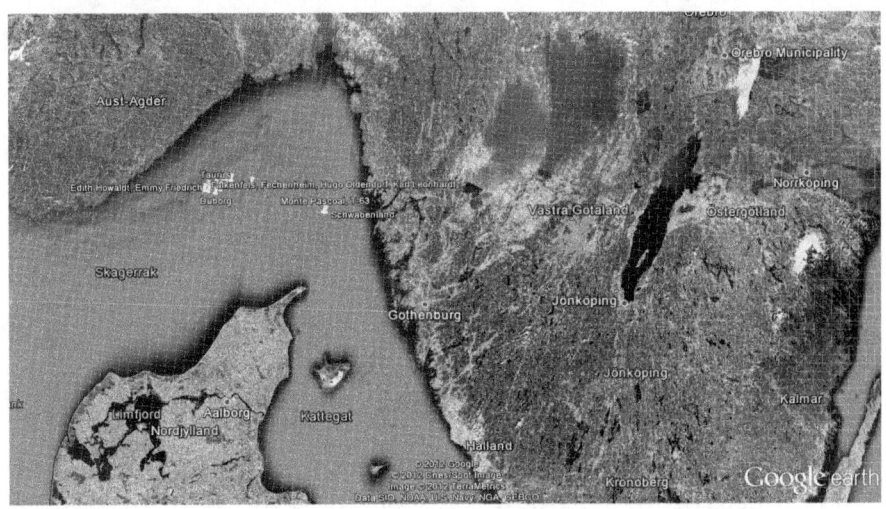

Figure 82
Map 1 of the 32 UK Ships Scuttled in the Skagerrak and
Kattegat, 1945-47

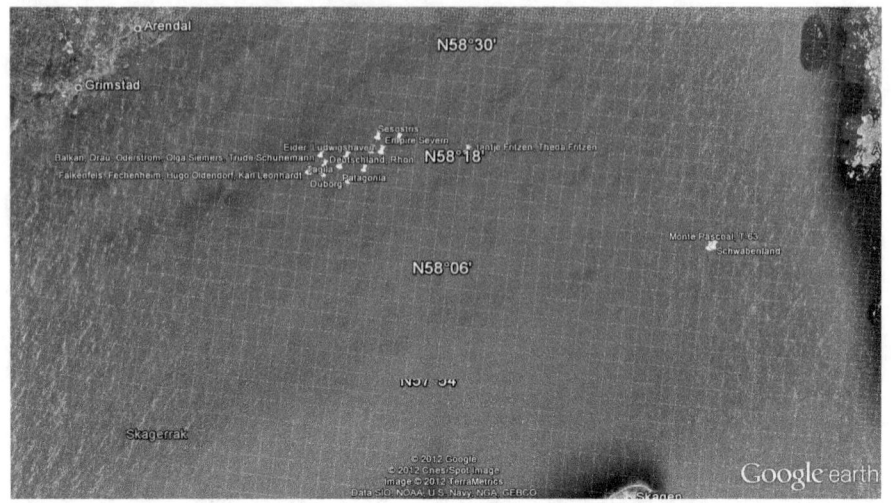

Figure 83
Map 2 of the 32 UK Ships Scuttled in the Skagerrak and
Kattegat, 1945-47

With specific regard to chemical munitions, the July 1945 report of the British Army of the Rhine (BAOR) noted that "it is believed that all major depots have now been discovered, the contents of the small dumps have been removed to the larger dumps (all of which are guarded), and there is a total of approximately 111,000 tons of enemy CW munitions in the British Zone." Regarding disposal options, the BAOR report also noted that considerable thought had been given to the problem of finding the most economical and acceptable method of disposal. Of the three methods considered (dumping at sea in deep water, interring in deep and flooded mines, or industrial break-down), sea disposal was considered necessary for the bulk of the munitions in question.

The Continental Ammunition Dumping Committee (CADC) held its first meeting on July 15, 1945 at HQ, 21st Army Group. The CADC was established by the

Service authorities in London to centrally control the task of ammunition disposal and in particular to organize the dumping at sea of captured enemy ammunition not required for research or to meet the requirements of the London Munitions Assignment Board (LMAB)[98]. (The LMAB was established to allocate specific items of German munitions, usually for research purposes, among the various members of the United Nations). Although there was close liaison with the American forces, and that information was passed to the Norwegian, Danish, and Dutch authorities, the CADC was not an inter-Allied body. It was a British organization established solely to deal with ammunition in the British Zone; stocks under British control in Belgium, the Netherlands, and Denmark; and certain stocks of surplus British ammunition which it was desired to dump.

Note that the CADC was sub-divided into the CADC (Germany), which assumed responsibility for operations on the continent and the CADC (UK) which handled the disposal of surplus British CW. Both the CADC (Germany) and the CADC (UK) reported to the London-based Continental Movements and Shipping Committee. According to the MoD, there is no evidence of any discussions of the CADC's program with the Russian authorities.[99]

It was agreed that the Army, through the 21st Army Group, would prepare the ammunition dumping program, including the program for CW munitions disposal. Note that this program only concerned dumping where the use of scuttled ships was concerned; the CADC was not responsible for local arrangements for the use of small craft such as lighters for the dumping of loose conventional ammunition. It was also decided at this meeting that "suitable" sites for the disposal of mustard gas (H) munitions included an area off Stavanger,[100] Norway at a depth of 100 fathoms (only 183 meters) and in the Kattegat[101]. No decision, however, could be made without further clearance from London.[102]

The following is excerpted from the 13th progress report of the Control Commission for Germany (British Element) covering the period 1-31 July, 1945 provided the following new instructions concerning CW disposal:

"The dumping of CW ammunition and stores has been the subject of separate consideration. The Ministry of Agriculture and Fisheries has agreed to dumping in waters of not less than 300 fathoms and an area has been established in the Skagerrak. It is proposed to make use of certain hulks which are beyond repair so as to economize in shipping; the hulks to be scuttled with their cargo of 'C.W.' material."[103]

Danish records reveal, however, the British marine attaché simply announced to the Danish marine ministry the intention of dropping large amounts of chemical weapons in international waters 15-20 nautical miles south of the Norwegian coast between 8°45' and 10°40' East.[104] The depth in that area is at least 450 meters (hereafter referred to as the "larger area").

The second meeting of the CADC took place on 2 August 1945 with the membership somewhat expanded, now including a representative from the U.S. Army's port command at Bremen. The details of the proposed dump sites off Norway and Denmark were to be passed to the Committee by the Naval Commander-in-Chief Germany so that they could be relayed to the Admiralty and the

Ministry of Agriculture and Fisheries. The area in the Skagerrak was considered especially suitable for CW dumping, as the depth of water there was more than 300 fathoms.[105]

On 20 August 1945, the Norwegian Directorate of Fishing objected to the larger area, but agreed to a smaller one 21-24 nautical miles south of the coastline between 9°27' and 9°40' east. The depth in that area is greater than 650 meters.

The official message to the CADC on August 20, 1945 read:

> From: Searail 7
>
> To: Chairman, Continental Ammunition Dumping Committee
>
> "Following for quartermaster general Norwegian Army begins ref No J nr 3758/45 JR/SK. The Director of Fisheries maintains that the ammunition ought to be taken out and dumped outside the continental platform either in the Atlantic Ocean or north of the North Sea alternatively in Skagerrak within an area enclosed by lines joining the following positions:
>
> A. 58 DEGS 14 MINS N 09 DEGS 27 MINS E
> B. 58 DEGS 16 MINS N 09 DEGS 27 MINS E
> C. 58 DEGS 19 MINS N 09 DEGS 40 MINS E
> D. 58 DEGS 17 MINS N 09 DEGS 40 MINS E.[106]
>
> Should this area be chosen, the area must be marked with buoys beforehand. The Ministry cannot agree to the dumping in any other place in Norwegian waters than those mentioned by the Director of Fisheries."

At the same time, "Store Nordiske Telegrafselskap" objected to the proposed dumping `because the main telephone cable between Norway and Hirtshals, Denmark went through the so-called larger area. Norwegian authorities also stated that all ammunition dumped should also be stored in ships.

Figure 84 provides the first details of the British Army's CW Ammunition Dumping Program. Depot Numbers and Locations, Tonnage, Loading Ports, and Allocated Shipping were provided in Appendix A to the minutes of the CADC Meeting #3, 16 August 1945.

DRAFT

C.W. AMMUNITION DUMPING PROGRAMME

Serial	Depot No.	Location	Map Ref	Corps Dist	Tonnage	Loading Port	Shipping Allocated
1	50	MOLLN	S 9606	8	3900	SCHLUTUP	CLARA RUSS (2000) PILLAU
2	109	LEHRE	Y 0318	30	8000		(1800) TRITON (2006)
3	35	SCHNEVERDEN	S 3405	30	700		NEPTUNE (2000) LOUISE SCHROEDER (1800) ALWINEROSS (1500) TRUDE SCHUNEMAN (1500) = 12,600
					12600		12600 @ 350 tpd = 36 days
4	24	MUNSTERLAGER	X 5890	30	16200	FLENSBURG	DRAU (8000) EDITH HOWALDT
5	29	CELLE	X599491	30	4500		(3000) BALKAN (3500) LOTTE
6	79	HANNIGSEN	X610350	30	600		(3500) DORA OLDENDORF (3500) = 21,500
					21300		21300 @ 450 tpd = 48 days
7	90	SENNELAGER	B 6853	1	3500	EMDEN	DUBORG (3500) KARL LEONHARDT
8	18	ESPELKAMPF	W 6020	1	11200		(8000) HELIOS (4000) OLGA SIEMERS (5000) FALKENFELS
9	22	RAHDEN	W505478	30	3600		(10000) EMMY FRIEDRICH
10	33	ORREL	X644880	30	13700		(8000) = 38500
11	78	BODENTEICH	X 9876	30	6500		
					38500		38500 @ 400 tpd = 97 days
12	38	WALSRODE	X253710	30	5500	NORDHAFEN	VICTORIA (8000) PATAGONIA
13	74	ILSTER	X 5393	30	8000		(8000) DER NEUTOHER (6000) ᵖJAN WELLENS (7000)
14	33	ORREL	X644880	30	19900		ˣSTETTIN (6000) JAN WELLENS
15	93	DIEKHULSEN	C 5092	30	3800		(3200) = 38200
16	79	HANNIGSEN	X610350	30	1000		
					38200		38200 @ 400 tpd = 96 days
				Grand total	110,600		ᵖJAN WELLENS does one voyage @ 7000 tons to dump and scuttles on last voyage @ 3200 tons
							ˣActs as accommodation ship to bring back crews of ships from scuttling area. On last voyage she lands to 6000 tons and is scuttled; other shipping provided to bring back crews.

Figure 84
C.W. Ammunition Dumping Programme
16 August 1945

An update to the 16 August "C.W. Ammunition
Loading Program" was provided in the appendices to the
minutes of the CADC's 4th Meeting held on 29 August
1945: "As of August 1945, 110,643 tons of chemical

warfare ammunition were scheduled for sea disposal. The Royal Navy is responsible for all arrangements for the scuttling of ships."

CW Ammunition Loading Programme

Serial	Depot	Location	Map Ref	Corps Dist	Tonnage	Loading Port	Ship	Quantity	Date Ready Loading	Start Loading	Complete Loading	ETA KIEL BAY
1	50	MÖLLN	S.9606	8	3900	SCHLUTUP	PILLAU	1800	5/9	5/9	9/9	14/9
2	109	LEHEE	Y.0348	30	8000		TRITON	2000	7/9	10/9	15/9	17/9
3	36	SCHNEVERDEN	S.3405	30	700		TRUDE	1800	12/9	15/9	15/9	21/9
							SCHÜNEMAN	3500	17/9	20/9	29/9	1/10
							BALKAN	1500	1/10			
							ALWINEROSS	2000				
					12600		CLARA RUSS	12600				
4	24	MÜNSTERLAGER	X.5890	30	16200	FLENSBURG	DRAU	8000	5/9	5/9	22/9	27/9
5	29	OERLE	X599491	30	4500		HELIOS	4000	22/9	22/9	30/9	5/10
6	79	HANNUSSEN	X610350	30	800		EDITH HOWALDT	3000	1/10	1/10	7/10	9/10
							DORA OLDENDORF	3000				
					21500		LOTTE	3500	21500			

Figure 85
CW Ammunition Loading Programme (Part 1)
29 August 1945

Depot	Location	Map Ref	Corps Dist	Tonnage	Loading Port	Ship	Quantity	Date Ready Loading	Start Loading	Complete Loading	ETA KIEL BAY
90	SENNELAGER	B.6853	1		BREMEN	LOUISE SCHRODER	1800	1/9	1/9	5/9	9/9
18	ESPELKAMP	W.6020	1	5500		DUBORG	3500	3/9	5/9	13/9	17/9
22	REHDEN	W.505478	30	11200		KARL LEONHARDT	8000	10/9	14/9	3/10	7/10
33	ORREL	X.644880	30	3600		PATAGONIA	8000	1/10	4/10	23/10	27/10
78	BODENTEICH	X.9876	30	12700		OLGA SIEMERS	5000				
						F.LKENFELS	10000				
						MONTE PASCOAL	1200				
				37500			37500				
38	WALSRODE	X.253710	30	5500	KIEL (NORDILFEN)	EMMY FRIEDRICH	8000	5/9	5/9	24/9	24/9
74	ILSTER	X.5393	30	8000		VICTORIA	8000	25/9	27/9	14/10	14/10
33	ORREL	X.644880	30	20900		JAN WELLENS	7000	15/10	15/10		
93	DIEKHOLSEN	0.5092	30	3800		DER DEUTCHER	6000				
						HESIPT	2200				
				38200			38200				
79	HANNIGSEN	X.610350	30	800		JAN WELLENS	7000	Shipping not yet allocated			

Figure 86
CW Ammunition Loading Programme (Part 2)
29 August 1945

By 31 August, 34 ships had been allocated and loading was scheduled to start during the first week of September 1945 at Kiel (Nordhafen), Emden, Schlutup, and Flensburg.[107]

Port Information:

Kiel:
http://www.port-of-kiel.de
World Port Source – Port Detail and Satellite Map:
http://www.worldportsource.com/ports/DEU_Port_o f_Kiel_356.php

Emden:
http://www.seaport-emden.de/cms/
World Port Source – Port Detail and Satellite Map:
http://www.worldportsource.com/ports/DEU_Port_o f_Emden_1253.php

Schlutup / Lubeck:
http://www.lhg-online.de/Schlutup.30+M52087573ab0.0.html
World Port Source – Port Detail and Satellite Map:
http://www.worldportsource.com/ports/DEU_Port_o f_Lubeck_386.php

Flensburg:
http://www.flensburg-tourismus.de/
World Port Source – Port Detail and Satellite Map:
http://www.worldportsource.com/ports/DEU_Port_o f_Flensburg_2790.php

Skagerrak Chemical Weapons Convoy #1 (CW1) – 4 October 1945

On 4 October 1945, the first Skagerrak chemical weapons convoy (CW1) was scuttled. According to the Ministry of Defence's 1993 "Report on the Sea Dumping of CW by the U.K. in the Skagerrak Waters Post WWII", five ships (the Duborg, Louise Schroder, Patagonia, Pillau, and the Triton), with a combined total of 17,100 tons of chemical munitions, were sent to the bottom. The second convoy was scheduled to sail on 11 October.

Vessel Name	CW Cargo (Tons)
Duborg	3,500
Louise Schroder	1,800
Patagonia	8,000
Pillau	1,800
Triton	2,000
CONVOY TOTAL:	17,100

Figure 87
Skagerrak Chemical Weapons Convoy #1 (CW1), 4 Oct 45
Vessels and CW Tonnage

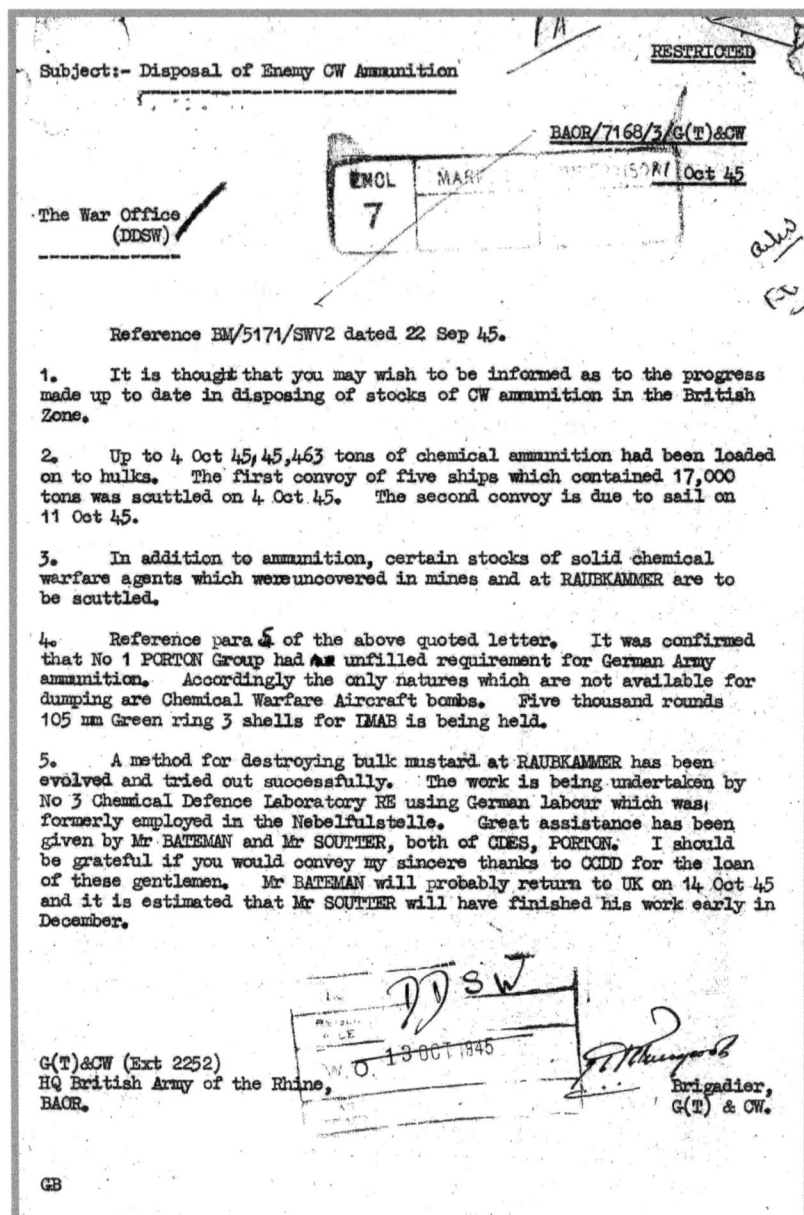

Subject:- Disposal of Enemy CW Ammunition

RESTRICTED

BAOR/7168/3/G(T)&CW

ENCL
7

MAR

Oct 45

The War Office
(DDSW)

Reference BM/5171/SWV2 dated 22 Sep 45.

1. It is thought that you may wish to be informed as to the progress made up to date in disposing of stocks of CW ammunition in the British Zone.

2. Up to 4 Oct 45, 45,463 tons of chemical ammunition had been loaded on to hulks. The first convoy of five ships which contained 17,000 tons was scuttled on 4 Oct 45. The second convoy is due to sail on 11 Oct 45.

3. In addition to ammunition, certain stocks of solid chemical warfare agents which were uncovered in mines and at RAUBKAMMER are to be scuttled.

4. Reference para 5 of the above quoted letter. It was confirmed that No 1 PORTON Group had an unfilled requirement for German Army ammunition. Accordingly the only natures which are not available for dumping are Chemical Warfare Aircraft bombs. Five thousand rounds 105 mm Green ring 3 shells for IMAB is being held.

5. A method for destroying bulk mustard at RAUBKAMMER has been evolved and tried out successfully. The work is being undertaken by No 3 Chemical Defence Laboratory RE using German labour which was formerly employed in the Nebelfulstelle. Great assistance has been given by Mr BATEMAN and Mr SOUTTER, both of CDES, PORTON. I should be grateful if you would convey my sincere thanks to CDDD for the loan of these gentlemen. Mr BATEMAN will probably return to UK on 14 Oct 45 and it is estimated that Mr SOUTTER will have finished his work early in December.

DDSW

13 OCT 1945

G(T)&CW (Ext 2252)
HQ British Army of the Rhine,
BAOR.

Brigadier,
G(T) & CW.

GB

Figure 88
Declassified RESTRICTED 11 October 1945 Message
Disposal of Enemy CW Ammunition

Skagerrak Chemical Weapons Convoy #2 – 17 October 1945

The second Skagerrak chemical weapons convoy was sunk on 17 October 1945. Seven hulks were scuttled (the Balkan, Drau, Edith Howaldt, Emmy Friedrich, Oderstrom, Olga Siemers, and the Trude Schunermann) with an estimated total of 32,215 tons of CW ammunition. The cumulative total for convoys 1 and 2 was 49,315 tons.

Vessel Name	CW Cargo (Tons)
Balkan	3,500
Drau	8,000
Edith Howaldt	3,000
Emmy Friedrich	8,000
Oderstrom	3,215 (est.)
Olga Siemers	5,000
Trude Schunermann	1,500
CONVOY TOTAL:	32,215
CUMULATIVE TOTAL:	49,315

Figure 89
Skagerrak Chemical Weapons Convoy #2 (CW2), 17 Oct 45
Vessels and CW Tonnage

Skagerrak Chemical Weapons Convoy #3 (CW3) – 17 November 1945

The third Skagerrak chemical weapons convoy was scuttled on 17 November 1945. Five ships (the Jantje Fritzen, Sesostris, Tagila, Taurus, and Theda Fritzen) with a combined total of 15,416 tons of CW were sunk.

Vessel Name	CW Cargo (Tons)
Jantje Fritzen	6,600
Sesostris	2,000
Tagila	2,600
Taurus	1,000
Theda Fritzen	3,216 (est.)
CONVOY TOTAL:	15,416
CUMULATIVE TOTAL:	64,731

Figure 90
Skagerrak Chemical Weapons Convoy #3 (CW3), 17 Nov 45
Vessels and CW Tonnage

Illustrating the inaccuracies that characterized the dumping operations, while the Ministry of Defence claims the Sesostris was scuttled at 58°15'N 09°30'E, the Norwegian Defence Research Establishment (NDRE), during a 1989 investigation, found the Sesostris with a sidescan sonar (with an accuracy of ± 20 meters) at 58°18' 31.5"N 09°41' 05.7"E. Additionally, the MoD has no record of the Theda Fritzen, confusing it with the Jantje Fritzen. In an article published in the Hamburger Abendblott on August 19, 1970, Friedrich Passehl, captain on board the Taurus on its final expedition, confirmed the sinking of the Theda Fritzen along with the other four ships in the third convoy.[108]

The MoD's 1993 "Report on the Sea Dumping of CW by the U.K. in the Skagerrak Waters Post WWII", also lists no cargo weight for the Oderstrom in CW2. As the cumulative total for the first three convoys, according to the 23 January 1946 SECRET message at Figure 91, is 64,731 tons – and given the combined total for 15 of the 17 ships scuttled in the first three convoys is 58,300 tons, the missing weight (6,431 tons) is estimated to be split between the Oderstrom in CW2 (3,215 tons) and the Theda Fritzen in CW3 (3,216 tons).

SECRET

Subject: Disposal of Enemy CW Stocks

BAOR/7168/3/G(T)&CW

23 Jan 46

The War Office,
(DSWV)

BAOR)DSWV (flag A)
-22A

Reference BAOR/7168/3/G(T)&CW dated 11 Oct 45.

1.   Up to 15 Jan 46 approx. 80,000 tons of enemy
CW ammunition had been moved to hulks for scuttling.
Three convoys totalling 64,731 tons have been
scuttled and a fourth convoy will sail shortly.

2.   28,478 tons available for dumping are now
held in RAUBKAMMER, MUNSTERLAGER and ORREL. Eventually,
all stocks of "frozen" material, ie. 8,303 tons of
Green Ring 3 A/C Bombs, 1050 x 250 kg White Ring Bombs
and 40 tons bulk CN will be held at ORREL pending
further instructions.

3.   All bulk stocks of liquid chemical warfare
agents amounting to some 4,000 tons have been destroyed
by burning. Destruction of smaller quantities of
chemical warfare material at RAUBKAMMER continues.

4.   In addition to German CW stocks, 24,000 tons
of US Army chemical warfare ammunition are beginning
to move to KIEL for inclusion in the present scuttling
programme.

B G S

G(T)&CW (Extn 2252)
HQ British Army of the RHINE

Figure 91
Declassified SECRET 23 January 1946 Message
Disposal of Enemy CW Stocks

In addition to citing the accurate total tonnage sea disposed in the first three convoys, the 23 January 1946 SECRET communication from HQ, BAOR to The War Office verifies that "in addition to German CW stocks, 24,000 tons of U.S. Army chemical warfare ammunition are beginning to move to Kiel for inclusion in the present scuttling programme".[109]

Between 16 March 1946 and 6 June 1947, 6 more chemical weapons convoys (CW4-CW10) comprised of a total of 15 vessels were scuttled in the Skagerrak. Unfortunately, MoD records are missing tonnages for 6 of the 15 ships (40%). Here are the known details:

	CONVOY	DATE	VESSEL NAME	CW Cargo (Tons)
18	CW4	16-Mar-46	Falkenfels	10,000
19			Fechenheim	Unknown
20			Hugo Oldendorf	Unknown
21			Karl Leonhardt	8,000
22	CW5	13-Jul-46	Freiburg	6,500
23			Gertrud Fritzen	4,500
24	CW6	8-Sep-46	Deutschland	2,121 shared
25			Rhon	
26	CW7	12-Oct-46	Eider (Sperrbrecher 36)	5,000
27			Empire Severn	4,700
28			Ludwigshaven	1,721
29	CW8	21-Dec-46	Monte Pascoal	Unknown
30			T-63	Unknown
31	CW9	17-May-47	Dessau	Unknown
32	CW10	6-Jun-47	Schwabenland	Unknown

Figure 92
Skagerrak Chemical Weapons Convoys 4-10 (CW4-CW10)
15 Vessels and CW Tonnage

In spite of difficulties in obtaining sufficient, suitable hulks, and with the severe weather of the winter of 1946-47 when some hulks were ice-bound, the last CW-loaded hulk, the Schwabenland, was scuttled in the Skagerrak on June 6, 1947, thus ending the ten-convoy, 32-ship,

610-day, sea-dumping program (October 4, 1945 - June 6, 1947).

The last BAOR Disarmament Progress Report to mention CW stocks is that of December 1947 which states that a total of 119,910 tons of CW munitions had been disposed of from British control.

Atlantic Disposal Operations

CW disposal operations in the Atlantic by the United Kingdom occurred in four phases over the period 2 July 1945 and September 1956.

According to official MoD records, a total of 24 ships loaded with chemical weapons were scuttled in the Atlantic between 2 July 1945 and September 1946. Twelve vessels were sunk north of Ireland and Scotland and twelve in the Bay of Biscay. (See Figures 93-96 below)

As stated in the UK Ministry of Defence's "Standard Reply to Enquiries Re Sea Dumping of Munitions", "Wreck dumping of CW in the Atlantic took place in four distinct phases. Phase 1 (1945-1948), 2 (1949-1951), 3 (1955-1956), and 4 (1956) involved the disposal of some 120,000 tons of mustard and phosgene-charged munitions."

Figures 97 and 98 display the MoD's official record of the scuttling date, location, vessel name, and disposal depth for the 24 ships scuttled in the Atlantic CW disposal operations.

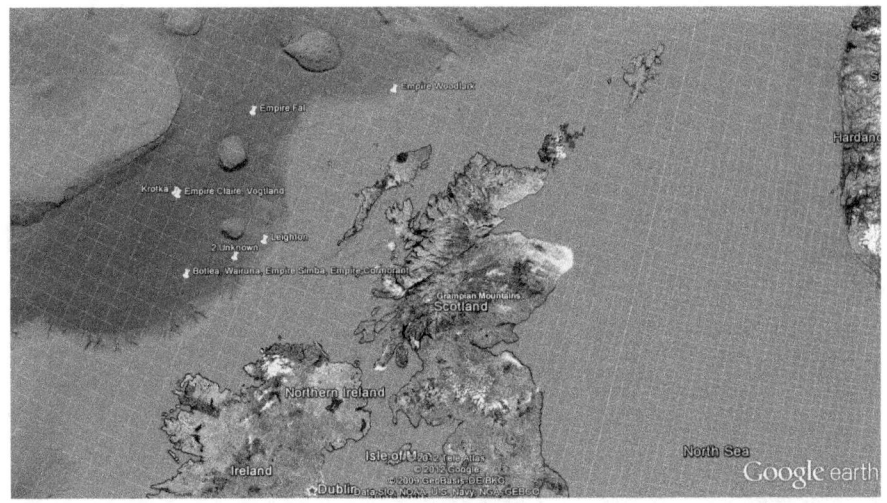

Figure 93
Map 1 of the 12 Vessels Scuttled
North of Ireland and Scotland

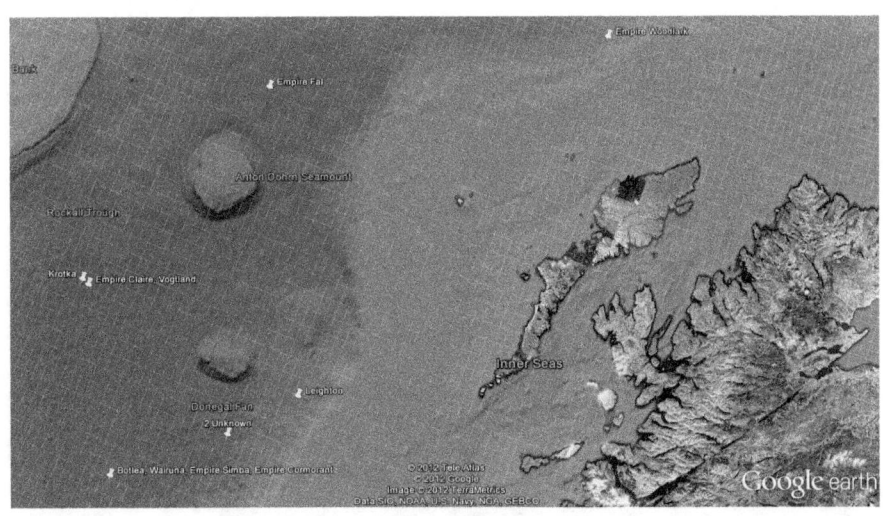

Figure 94
Map 2 of the 12 Vessels Scuttled
North of Ireland and Scotland

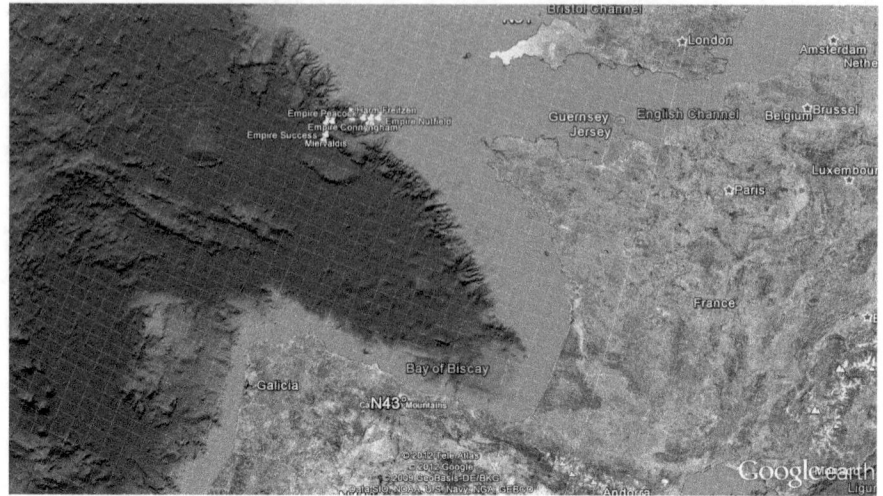

Figure 95
Map 1 of the 12 Vessels Scuttled
in the Bay of Biscay

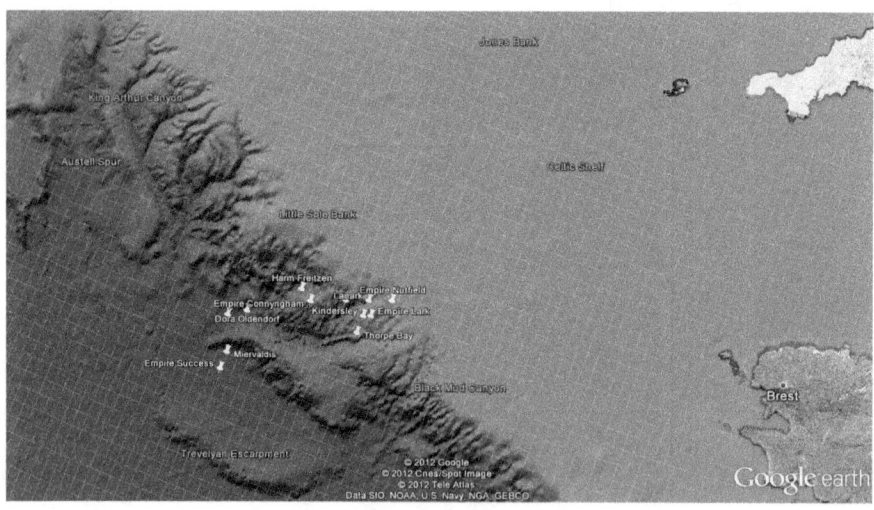

Figure 96
Map 2 of the 12 Vessels Scuttled
in the Bay of Biscay

Date	Location	Ship	Depth
Phase 1			
02.07.45	Lat 58 00,9 N Lon 11 00.0 W	Empire Fal	2000M
11.09.45	Lat 55 30.00 N Lon 11 00.00 W	Empire Simba	2500M
01.10.45	Lat 55 30.00 N Lon 11 00.00 W	Empire Cormorant	2500M
30.10.45	Lat 55 30.00 N Lon 11 00.00 W	Wairuna	2500M
30.12.45	Lat 55 30.00 N Lon 11 00.00 W	Botlea	2500M
25.08.46	Lat 47 57.00 N Lon 08 33.24 W	Empire Peacock	700-800M
03.09.46	Lat 48 03.00 N Lon 08 09.00 W	Empire Nutfield	500M
01.10.46	Lat 47 54.00 N Lon 08 21.00 W	Kindersley	1000M
02.11.46	Lat 59 00.00 N Lon 07 40.00 W	Empire Woodlark	800M
11.11.46	Lat 48 00.00 N Lon 08 21.00 W	Lanark	800-900M
05.02.47	Lat 47 40.00 N Lon 09 22.00 W	Dora Oldendorf	3500-4000M
27.07.47	Lat 47 55.00 N Lon 08 17.00 W	Empire Lark	750-800M
09.08.47	Lat 56 22.00 N Lon 09 27.00 W	Leighton	1300M
08.09.47	Lat 47 47.30 N Lon 08 21.00 W	Thorpe Bay	1500M
03.11.47	Lat 47 36.00 N Lon 09 31.00 W	Margo	4100M
01.03.48	Lat 47 55.00 N Lon 08 58.00 W	Harm Freitzen	2500M
22.08.48	Lat 47 16.30 N Lon 09 24.00 W	Empire Success	4200M
22.09.48	Lat 47 23.00 N Lon 09 24.00 W	Miervaldis	4000M

Figure 97
Phase 1, Atlantic CW Disposal Operations, 1945-1948[110]

Date	Location	Ship	Depth
Phase 2			
20.06.49	Lat 47 52.00 N Lon 08 51.00 W	Empire Connyngham	2000M
Phase 3 – Operation SANDCASTLE			
27.07.55	Lat 56 30.00 N Lon 12 00.00 W	Empire Claire	2500M
30.05.56	Lat 56 30.00 N Lon 12 00.00 W	Vogtland	2500M
23.07.56	Lat 56 31.00 N Lon 12 05.00 W	Krotka	2500M
Phase 4			
??.06.56	Lat 56 00 N Lon 10 00 W	Unknown	2000M
??.06-09.56	Lat 56 00 N	Unknown	2000M

Figure 98
Phases 2-4, Atlantic CW Disposal Operations, 1949-1956[111]

Phase 3 (Operation "Sandcastle") of the Atlantic CW sea disposal operations, which occurred from 27 July 1955 through 23 July 1956, "involved the disposal of 14,000 tons of Tabun-charged munitions confiscated from Germany after World War II. Additionally, 300 tons of arsenical compounds (in powder form and sealed in drums from a Ministry of Supply depot) and 3 tons of toxic seed dressings (in 50 containers from HM Norfolk Flax Establishment) were dumped during this operation."[112]

Operation Sandcastle (1955-1956)

Early in 1954, the problem of continued storage of approximately 70,737 Tabun (GA) bombs at Llandwrog was under consideration by senior Air Ministry staff. The ultimate solution, agreed in June 1954 by the Air Council and Service Ministers, was to dump the chemical weapons at a sufficiently distant and deep location in the North Atlantic. By November 1954, the following plan for Phase I of two phases of the Llandwrog operation, codenamed Operation Sandcastle, had been drafted by the Royal Air Force's No. 42 Group:

1. Transfer the chemical containers containing a total of 16,000 tabun bombs from the airfield (RAF Llandwrog) to the nearby beach at Fort Belan, North Wales.
2. Load the containers via roller conveyors on trestles into Army landing craft operated by No. 45 Water Company of the RASC.
3. Transport the containers by sea to Cairnryan, Scotland – a small and isolated military port five miles north of Stranraer.
4. Reload the bombs into old hulks.
5. Scuttle the hulks in three sites centered on position 56°30'N 12°05'W – approximately 120 miles to the northwest of northern Ireland at a depth of 6,000 feet.

In Phase II of Operation Sandcastle, the remaining stockpile of Llandwrog's unwanted munitions – approximately 54,737 bombs – would be sea disposed in the spring of 1956.

Phase I of the project was initiated on 11 January 1955 with loading trials conducted at the Fort Belan beachhead using landing craft from Cairnryan. Although the loading trials were successful, the operation was delayed until summer to avoid the risks of transporting the toxic material under winter conditions and to install gas detection devices and radio equipment to ensure constant communication with Llandwrog during all stages of the 18-hour journey between North Wales and Scotland. With the arrival of 157 airmen at the airfield in early June 1955 to carry out the various tasks involved with the project, Operation Sandcastle officially commenced.

The start of the dumping operation is recorded as follows: "On 13 June at 2:30pm, Landing Craft L408 beached adjacent to the roller conveyor line. Loading began at 6:52am on June 14th (1955) in pouring, driving rain. Feeding bombs from the stockpile to the ship proved no problem but stacking them initially posed many problems due to inexperienced personnel. It was noticeable as the day wore on, the stacking speed increased." L408 arrived safely at Cairnryan on June 15th with 400 boxed bombs. This was the first of 32 such journeys made by the six Operation Sandcastle landing craft during the summer of 1955.

Work continued with increasing efficiency – and in just over one month (by Sunday, July 24th) all 16,000 tabun bombs had been removed from Llandwrog, Wales, transported by sea to Cairnryan, Scotland and loaded into the decaying hulk of the 5,613 ton SS Empire Claire. The Empire Claire was scuttled at 10:00am, Wednesday, 27 July 1955 completing Phase I of Operation Sandcastle.

Initiating Phase II, during the winter months of 1955-56, approximately 54,737 tabun bombs were "de-tailed" in preparation for shipment by landing craft to Cairnryan. The first load of bombs arrived from Llandwrog at 1:35am, Saturday, 7 April 1956 on Landing Craft L406. By 27 May 1956 the 4,960 ton German freighter MV Vogtland had reached its capacity load of 28,737 bombs and on 28 May the hatches were battened down. At 9:00pm, Wednesday, 30 May 1956 the Vogtland was scuttled in the North Atlantic.

After a short period of leave for the Royal Air Force personnel, loading of the third and final ship in Operation Sandcastle, the 6,103 ton SS Kotka, began on 11 June 1956. In addition to the 26,000 remaining tabun-filled bombs from Llandwrog, the Kotka also had on board 330 tons of arsenical compounds in powder form received from the Minister of Supply. There were also 50 cases of toxic seed dressing from HM Norfolk Flax Establishment. Research indicates this "toxic seed dressing" may be related to the anthrax spores experiments carried out in the 1940s. According to official records, the last bomb of Operation Sandcastle was stowed in the shelter deck of No. 5 hold at 2:45pm, Saturday, 21 July 1956. The S.S. Kotka was scuttled in the North Atlantic at 12:50pm, Monday, 23 July 1956.

As revealed in the House of Commons Hansard Debates for 30 March 1995, "the vessels scuttled in Operation Sandcastle sank in water up to 2,000 (meters) deep some 80 miles north-west of the coast of Northern Ireland. No monitoring of these dump sites has been undertaken because of the depth of the water and also because the cargo, comprising the nerve agent Tabun, is destroyed by hydrolysis and rendered harmless in sea water."[113]

For reference, the behavior of individual chemical warfare agents (including Tabun) in seawater and their effects on the marine environment are rigorously addressed in Chapter 5.

Note also that in addition to the disposal of Tabun-filled bombs, Operation Sandcastle is alleged to have also included weapons containing cyanide, drums of both Tabun (GA) and Sarin (GB) nerve agents, and large quantities of mustard gas.[114]

The final phase (Phase 4) of the United Kingdom's Atlantic sea disposal operations occurred between June and September 1956. Two ships, of unknown names were scuttled. The first, according to the 1993 MoD report, was sunk near 56° 00' 00"N, 10° 00' 00"W at a depth of 2,000 meters. According to the 2002 OSPAR Commission report "Overview of Past Dumping at Sea of CW and Munitions in the OSPAR Maritime Area, the second ship was sunk at approximately 56° 22' 12"N, 09° 27' 00"W.

It is a concern that so little is known about the most recent UK disposal operations. At the MoD's own admission, "records for Phase 4 are incomplete and there is no clear evidence that the usual practice of sealing CW into containers and dumping it in scuttled merchant ships was followed during this phase."[115]

BOTTOM LINE: Including the disposal of excess stocks, at least 56 ships loaded with a total of at least 247,000 tons of chemical weapons were scuttled by the United Kingdom between 1945 and 1956 – 32 ships with 127,000 tons in the Skagerrak and 24 ships with 120,000 tons in the Atlantic.

3.2.3 Russia – Disposal Operations

Between 1946 and 1978, Russia dumped 356,872 tons of chemical weapons in the Baltic primarily in two charted regions. This specific number is cited in a 19 October 1989 document of the Central Committee, Communist Party of the Soviet Union (See Figure 103). Note this tonnage was loose dumped, as opposed to being scuttled in ships' hulls. An additional 112,523 tons of excess CW stocks were directed to be similarly disposed between 1989 and 1990. Although these disposal operations are corroborated by a 17 March 1993 publication in the St. Petersburg Chas Pik entitled "Chas Pik Investigation: Secrets of Chemical Warfare" which included an interview with Dr. of Chemical Sciences Lev Aleksandrovich Fedorov, official documentation detailing the actual disposal operations conducted in 1989-1990 has yet to be obtained.

Contrary to the 19 October 1989 directive from the Central Committee, Communist Party of the Soviet Union, the October 1993 Report of the Russian Federation, "Complex Analysis of the Hazard Related to the Captured German Chemical Weapon (sic) Dumped in the Baltic Sea" states only 70,500 tons of CW containing 12,035 tons of chemical warfare agents were dumped in two principal regions of the Baltic Sea in 1947-48.

Figure 99 displays the two primary Russian disposal regions in the Baltic Sea.

- Region 1: A polygonal area 75 km south of Gotland, Sweden, bounded by the following seven coordinates: 56°16'N 18°39'E, 56°16'N 18°51'E, 56°20'N 18°55'E, 56°20'N 19°31E, 56°07'N 19°15'E, 55°56'N 19°15'E, and 55°56'N 18°39'E.[116]

- Region 2: A rectangular area east north-east of Bornholm Island, Denmark, bounded by the following coordinates: 55°23'N 15°28'E, 55°23'N 15°55'E, 55°07'N 15°55'E, and 55°07'N 15°28'E.[117]

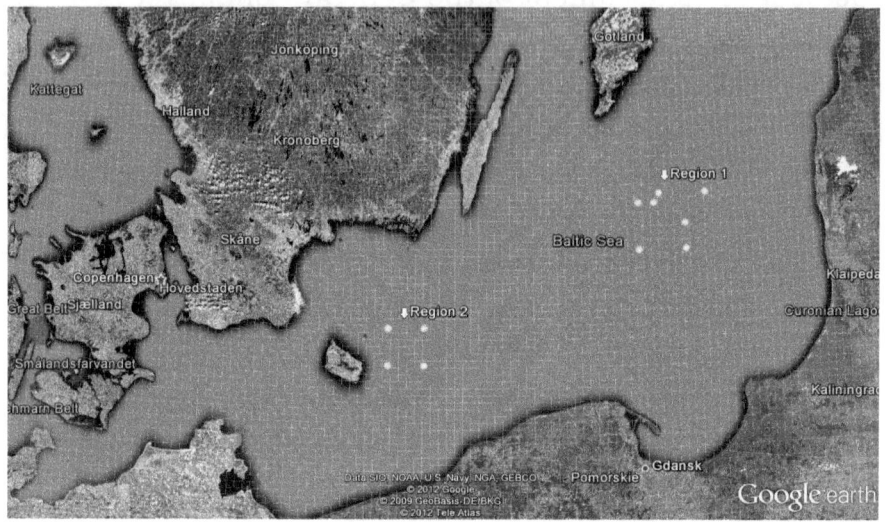

Figure 99
Primary Russian Disposal Regions in the Baltic Sea

Details of the 1947 Disposal Operation

"In the period from the end of the Second World War to the spring of 1947, the Soviet occupation forces collected war materiel originating from the German Wehrmacht in a depot outside (the German port city) Wolgast in Pommern.

Link: **http://en.wikipedia.org/wiki/Wolgast**

The depot in Wolgast was under Soviet military administration, and its leadership, watch crew, etc., comprised Soviet officers and soldiers. The storekeepers and similar personnel were German conscripts from the area around Wolgast who were ordered by the Soviet occupation troops to carry out the work in question. The materiel from the depot was loaded onto eight German merchant ships which the Soviet authorities commanded during the period 19 May 1947 to 10 January 1948, and was then thrown into the Baltic Sea.

The ships, whose home ports were in the British and American zones of occupation, were ordered to Wolgast where, under Soviet command, they were loaded and sailed to the dumping sites near Gotland (Region 1) and Bornholm (Region 2). The greatest quantities of ammunition were thrown overboard near Bornholm. [The 1993 National Report of the Russian Federation confirms this. Of the 608,482 pieces of CW materiel disposed in the Baltic in 1947, 560,090 pieces (92%) were dumped in Region 2.] After a few cargo loads were dumped near Gotland (Sweden) (Region 1), the dumping in this area was discontinued since the distance to the dumping sites near Bornholm was considerably shorter.

Figure 100 provides information on the eight ships used in the dumping operations east of Bornholm Island, Denmark during the period 19 May 1947 and 10 January 1948. These eight vessels alone made a total of 90 separate journeys to loose-dump chemical weapons in the Baltic.

Table. 4: List of the ships involved in the dumping operations east of Bornholm

No.	Ship	GRT	Dead-weight	Charter period; no. of journies
1.	Christian	795	1000	18.08.47-23.12.47; 16
2.	Jupiter	560	762	29.10.47-10.01.48; 8
3.	Venus	628	833	25.10.47-30.12.47; 17
4.	Rhein	800	1200	22.10.47-02.01.48; 6
5.	Odermünde	577	920	04.09.47-22.12.47; 14
6.	Brake	617	900	20.08.47-31.12.47; 13
7.	Elbing IV	327	500	19.05.47-22.09.47; 8
8.	Elbing VIII	590	950	8

Figure 100
List of Ships Involved in the Dumping Operations
East of Bornholm Island, Denmark
19 May 1947 – 10 Jan 1948[118]

On 30 December 1947, the Soviet military authorities announced that the dumping had been discontinued (although the final dumping operation was conducted on 10 January 1948 by the "Jupiter") (See #2 in Figure 100). Simultaneously, it was declared that the dumping had occurred in an area surrounding the coordinates 55°20'7" North and 15°37'8" East."[119]

The October 1993 Report also provides exact details on the 608,482 individual pieces of CW material (bombs, shells, mines, grenades, cans, drums, barrels, and containers) which comprised the 70,500 tons. Figure 101 presents this information by piece type, agent, and disposal region.[120]

TYPE OF CW		AGENTS				
Region and Subtotals						
		H	Arsinöl	DM	CN	Other	TOTAL
Aircraft Bombs	1	5,690	721	639	414	0	7,464
	2	65,779	8,338	7,388	4,785	0	86,290
		71,469	9,059	8,027	5,199	0	93,754
Artillery Shells	1	26,205	0	2,564	4,300	0	33,069
	2	302,926	0	29,639	49,702	0	382,267
		329,131	0	32,203	54,002	0	415,336
HE Bombs	1	2,720	0	0	0	0	2,720
	2	31,442	0	0	0	0	31,442
		34,162	0	0	0	0	34,162
Mines	1	830	0	0	0	0	830
	2	9,590	0	0	0	0	9,590
		10,420	0	0	0	0	10,420
Barrels	1	42	124	0	0	48	214
	2	487	1,434	693	0	552	3,166
		529	1,558	693	0	600	3,380
Aerial Smoke Grenades	1	0	0	2,790	0	0	2,790
	2	0	0	32,250	0	0	32,250
		0	0	35,040	0	0	35,040
Containers	1	0	80	0	0	0	80
	2	0	924	0	0	0	924
		0	1,004	0	0	0	1,004
Drums	1	0	0	599	0	0	599
	2	0	0	6,927			6,927
		0	0	7,526	0	0	7,526
Cans	1	0	0	0	0	626	626
	2	0	0	0	0	7,234	7,234
		0	0	0	0	7,860	7,860
TOTAL Region 1:		35,487	925	6,592	4,714	674	48,392
TOTAL Region 2:		410,224	10,696	76,897	54,487	7,786	560,090
GRAND TOTAL:		445,711	11,621	83,489	59,201	8,460	608,482

Figure 101
Total Number of Chemical Weapons Disposed
in the Baltic Sea by Russia, 1947

Disposals of Excess National Stocks

As highlighted earlier, although the 1993 Report of the Russian Federation states that 70,500 tons of CW were found in the Russian Zone and disposed in the Baltic Sea in 1947, another key source indicates a significant amount of excess national stocks may have been sea-disposed up until 1990.

The following is a summary of a document entitled "Information to Comrade L.N. Zaikov's Note, signed on 19 October 1989 by V. Maidannikov, Organizational Chief and I. Pismennik, Defense Work Department Chief, Central Committee, Communist Party of the Soviet Union. This document likely became available as a result of President Gorbachev's policy of glasnost (literally "openness").[121] Mikhail Sergeyevich Gorbachev served as Head of State of the Soviet Union (and as President of the Soviet Union from 15 March 1990) from 1 October 1988 to 25 December 1991. He was awarded the Nobel Peace Prize in 1990.

- **356,872 tons of chemical ammunitions and containers with chemical agents were dumped in the Baltic between 1946 and 1978. This tonnage included:**
  - 408,565 projectiles: 75mm and 150 mm, mustard gas (H)
  - 71,469 aircraft bombs, 250 kg, mustard gas (H) (This number precisely matches the 1993 Russian Report data for the 1947 disposal.)

- o 17,543 aircraft bombs, adamsite (DM) and arsinöl (According to the 1993 Russian Report, 17,086 were included in the 1947 disposal.)
- o 1,564 containers, 1500kg, mustard gas (H)
- o 10,420 chemical mines, 100mm (This number precisely matches the 1993 Russian Report data for the 1947 disposal.)
- o 7,295 barrels with chemical grenades
- o 7,860 barrels with "Cyclone" gas (This number precisely matches the 1993 Russian Report data for the 1947 disposal.)
- o 189,000kg of cyanic salt

- **112,523 tons of "outdated chemical ammunitions produced in 1954-1962" were directed to be disposed in the Baltic between 1989 and 1990.**

CHEMICAL WEAPONS IN THE BALTIC SEA

Secret things always come to light. Ministers and generals argued furiously that the USSR had submerged captured toxic agents by Bornholm and Liepaya only in 1947. But an expedition by Ekobaros Program developed by St. Petersburg Association "Okeanotehnika", approved by all the ministries of the Russian Government, was frozen by academician A. Kuntsevich, Chairman of the Committee on Conventional problems of Chemical and Bacteriological Weapons under President of Russia. By the very academician - general of chemical troops awarded Lenin Prize for development of new binary weapons from the hands of Nobel Peace Prize winner M. Gorbachyov, when the latter was President of the USSR. Now he is at the head of the International Green Cross. A. Kuntsevich froze Ekobaros Program, because these investigations became an international scandal - Soviet bombs and containers with more lethal toxic agents would be found over German ones. From the document we are publishing below it is obvious that the military-industrial complex was going to conduct new more large-scale submersions in 1989-1990.

rp-12/66

Information to comrade L.N. Zaikov's note

From 1946 till 1978 by means of units and sub-units of the Red-Banner-holding Baltic Navy 356872 tons of chemical ammunitions and containers with chemical agents were submerged in the area of the Baltic Sea. Among them: -

- 408565 projectiles of 75 and 150 mm calibre charged with yperite,
- 71469 aircraft bombs weighing 250 kg, charged with yperite,
- 17 543 aircraft bombs, charged with adamsite and arsenile,
- 1564 containers weighing 1500 kg, loaded with yperite,

Figure 102a
19 October 1989 Directive from the Central Committee,
Communist Party of the Soviet Union (CPSU) (Page 1 of 2)

2

- 10420 chemical mines of 100 mm calibre,
- 7295 barrels with chemical grenades,
- 7860 barrels with gas "Cyclone",
- 189000 kg of cianic salt.

Ammunitions were submerged at a depth from 100 to 150 m in two main points - 70 km from the Liepaya Naval Base as well as in the region of the island of bornholm. Exact coordinates of the submersion points wer plotted in nautical charts of the Navy confirmed for usage by Order of the Naval Commander-in-Chief dated 12.07.1985. Concerning prediction of specialists from the Main Naval Administration of Navigation and Am- munition, NPO-4 "Typhoon" of the USSR Goskomgidromet, map of chemical weapons submersions compiled by the Headquarters of the Red-Banner- holding Baltic Navy and analysis of samples lifted in 1983-1984, I find it reasonable to conduct additional submersions of outdated chemical ammunitions produced in 1954-1962 at the points of old submersions within 1989-1990. The total weight of ammunitions to be utilized is 112523 tons. Submersions are supposed to be made by means of ships from the Kaliningrad and Liepaya Naval Bases. Also I find it reasonable to instruct the headquarters of chemical protection troops to check the state of ammunitions adopted by units and sub-units of chemical protection and civil defence troops. Nominate comrade A. Kuntsevich responsible. Prepare escort ships and vessels of the Kaliningrad Naval Base for checking sailing and navigation conditions in ammunition sub- mersion points.

Prepare reference-plan of submerging chemical agents and ammu- nitions in the area of the Baltic Sea for a report to the Central Committee of the Communist Part-y of the Soviet Union 14.11.1989 at the latest.

V. Maidannikov, CC CPSU Organisational
Department Chief
I. Pismennik, CC CPSU Defence Work
Department Chief

October 19, 1989

3 copies.
Comp. D.S. Arhipova
rp -12/66

Editor's note: commentary to the article by Viktor Teryoshkin will be published in the next issue of "Chas Pik"

Figure 102b
19 October 1989 Directive from the Central Committee,
Communist Party of the Soviet Union (CPSU) (Page 2 of 2)

It is noted that "the exact coordinates of the submersion points were plotted in nautical charts of the Navy, confirmed for usage by Order of the Naval Commander-in-Chief dated 12.07.1985."

The Editor's note at the bottom of the document indicates the next issue of the St. Petersburg Chas Pik will include a commentary on the document by author Viktor Teryoshkin. The following text was, in fact, included in the 17 March 1993 issue of the St. Petersburg Chas Pik by Viktor Tereshkin in an article entitled "Chas Pik Investigation: Secrets of Chemical Warfare":

"The first wave of the 'salvaging' of the military toxic agents came after the end of the war. (The Baltic disposals in 1947 cited in the 1993 Russian Report)

The second wave arrived in the 1960s. This one, the one of the 'thaw' was simply a panic. Chapayevsk received a telegram from Chukhnov, then chief of the chemical forces, stating that they were to destroy all stocks of toxic agents at once! Again next to the industrial site they opened up the trenches, poured the mustard gas in there, hurriedly sprinkled something on top of it, and covered it up with earth. They proceeded differently with the lewisite bombs. They loaded them on 50 to 60 convoys and sent them directly to the Arctic Ocean.

The last wave – the big one – came at the end of the 1980s. The talks on chemical disarmament were coming to an end and it was clear that our stocks greatly exceeded those of the Americans. This was the source of the figure of 112,523 tons of munitions that the CPSU Central Committee planned to dump near Bornholm (Denmark) and Liepaja (Latvia) using ships from the Kaliningrad and Liepaja naval bases."[122]

BOTTOM LINE: Russia disposed of between 356,872 and 469,395 tons of chemical weapons in the Baltic Sea between 1946 and 1990.

3.2.4 Germany – Disposal Operations

In the last days of the war, as cited in the May 1993 German report "Chemical Munitions in the Southern and Western Baltic Sea" by the Federal Ministry of Transport's Maritime and Hydrographic Agency, "the German Navy sank two ships containing about 69,000 tabun shells and a further 5,000 tons of chemical munitions [phosgene (CG) and tabun (GA)] in the area south of the Baltic's Little Belt.[123]

http://en.wikipedia.org/wiki/Little_Belt

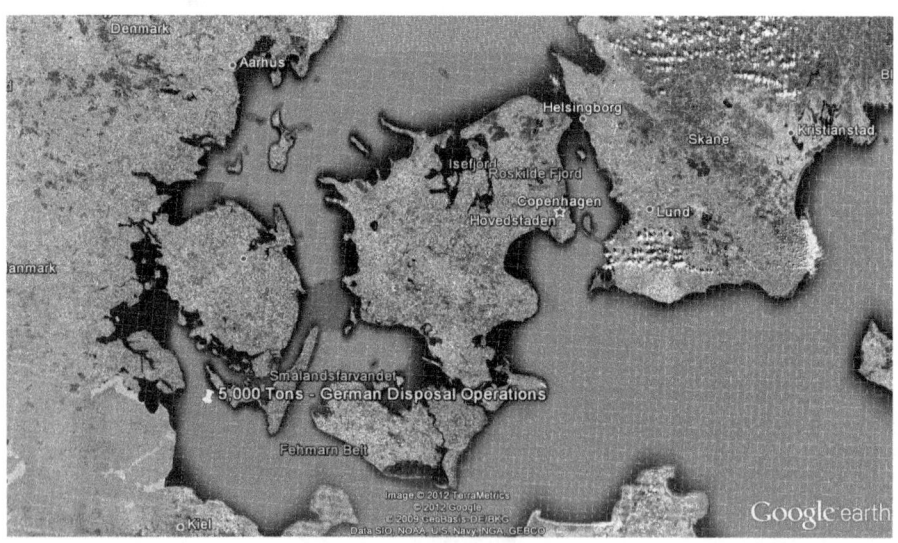

Figure 103
Map of German Disposal Site
in the Little Belt, Baltic Sea

The area in the Little Belt is situated between 54°47'-54°50' North and 10°08' - 10°15' east. The water is around 30 meters deep and the sea bed is covered with mud up to 8 meters thick.[124]

This disposal was obviously not charted, as:

"In 1954, German wreck salvagers "noticed" that there were two German wrecks laden with nerve gas (tabun shells) located at the Little Belt's southerly approach seven nautical miles south-east of Pöls Huk, immediately outside the Danish territorial waters. The Germans informed the Danish authorities of this "discovery" and of the measures taken to preclude potential risks associated with the wrecks' dangerous cargoes.

During the investigation by the Germans into the more specific circumstances surrounding the dumping, it was noted these were carried out by the German authorities shortly before the end of the Second World War.

The dumping occurred when two German ships [the larger of which was a vessel of approximately 1,000 gross register tons (GRT)] laden with nerve gas shells (10.5 cm and 15 cm caliber) and a large quantity of gunpowder, were sunk. The ships were located at 54°48'22" North and 10°13'22" East.

In 1959-60, the German authorities brought the ships to the surface, and the detonators and explosive charges were removed. The nerve gas shells were then encapsulated in concrete and dumped in the Bay of Biscay in March 1960.[125]

http://en.wikipedia.org/wiki/Bay_of_Biscay

Tables provided in the May 1993 German report give a more accurate location of the 1959-60 recovery of the 69,000 tabun shells: 54°48'7" North, 10°13"5' East.

Geographically, it is a significant distance to travel from the Baltic's Little Belt to the Bay of Biscay. It would be interesting to know 1) Germany's decision-making process to re-dump the 69,000 tabun-filled shells off the French coast and 2) the precise coordinates of the new dumpsite.

Around 5,000 tons of bombs and shells containing phosgene and tabun were thus left in the (Little Belt) area."[126] This number is included in the summary chart of Total CW Tonnage Sea Disposed in European Waters – by Nation, at Figure 55.

The May 1985 Report of the Danish National Environmental Protection Agency gives a more amplified accounting of the Little Belt disposal operations:

"The dumping south of the Little Belt...consisted of the sinking of two German ships loaded with nerve gas (tabun shells) and of 5,000 tons of chemical weapons (bombs and shells) and approximately 600 tons of V-1 rockets equipped with conventional explosives, all of which were thrown overboard.

The 5,000 tons of chemical weapons and approximately 600 tons of V-1 rockets were dumped by three German ships (the "Taurus", the "Karoline" and the "Marie Louise") in a total of seven expeditions.

The German authorities raised the two ships in 1959-60, but the other chemical weapons and the V-1 rockets are still to be found in the dumping area, where the water depth is 20-30 meters.

A discovery made by the German authorities in the spring of 1971 revealed that the corroded bombs containing the choking gas phosgene (CG), which, in contrast to several other types of poison gas, breaks down rapidly on contact with water. The artillery shells were, however, intact and functional and contained the nerve gas tabun (GA).

The chemical weapons dumped must be assumed to be lying spread around the dumping site. This is because on five of the total of seven dumping expeditions, the dump ships were not anchored, and because the ammunition in certain cases was encased in crates which drifted on the surface of the water before they sank. On two of the dumping expeditions the ships were anchored and in one of the seven expeditions the ammunition was removed from the crates before being dumped."[127]

BOTTOM LINE: Germany disposed of 5,000 tons of chemical weapons in the Little Belt of the Baltic and re-disposed 69,000 Tabun (GA) nerve gas shells (originally dumped in the Little Belt) in the Bay of Biscay.

3.2.5 Additional Ships with Suspected CW Cargoes

Research indicates 22 additional vessels which were scuttled in the Skagerrak and Kattegat may also have contained CW cargoes. Official records have also revealed at least 11 other vessels which were scuttled in European waters after WWII possibly contained chemical weapons. Information currently available regarding these ships is listed in Chapter 4, Locations of the Scuttled Ships.

Figure 104
Map of the 22 Additional Ships Suspected
of Containing CW Cargoes

CHAPTER 4

LOCATIONS OF THE SCUTTLED SHIPS

4.1 Introduction

Research has shown that as many as 100 vessels with chemical weapon cargoes were scuttled in the Skagerrak, Kattegat, Norwegian Sea, Atlantic Ocean, and the Bay of Biscay after World War Two. Loose dumping of CW was carried out in the Baltic by the Russians, in the Baltic and the Bay of Biscay[128] by the Germans, and in the Mediterranean by the Americans. From official government records, 67 ships are known to have been laden with chemical weapons; 22 more are suspected to have sunk in the Skagerrak or Kattegat with CW; and an additional 11 vessels – which lack details on tonnage and disposal dates and locations – are cited in official documents and national reports.

4.2 Lists of the Scuttled Ships by Geographic Area

Figures 105 through 108 list the 67 known CW vessels by geographic area. If available, data on CW tonnage, date of scuttling, and location are provided. Figure 109 addresses the 22 suspected CW vessels sunk in the Skagerrak and Kattegat and Figure 110 lists the 11 additional ships which may have been included in CW disposal operations. An alphabetical listing of all 100 ships is at Appendix D.

	SKAGERRAK AND KATTEGAT STRAITS	Tonnage	Date Scuttled	Latitude	Longitude
1	ALCO BANNER	3,097	14-Jul-46	58 18' 07" N	09 36' 05" E
2	BALKAN	3,500	17-Oct-45	58 16' N	09 26' E
3	DESSAU	6,000	17-May-47	Unk	Unk
4	DEUTSCHLAND	1,061	08-Sep-46	58 17' N	09 36' E
5	DRAU	8,000	17-Oct-45	58 16' N est.	09 26' E est.
6	DUBORG	3,500	04-Oct-45	58 14.5' N	09 31' E
7	EDITH HOWALDT	3,000	17-Oct-45	58 15' N	09 30' E
8	EIDER (Sperrbrecher 36)	5,000	12-Oct-46	58 17' N	09 38' E
9	EMMY FRIEDRICH	8,000	17-Oct-45	58 14' N	09 27' E
10	EMPIRE SEVERN	4,700	12-Oct-46	58 18' 30" N	09 37' E
11	FALKENFELS	10,000	16-Mar-46	58 14' N est.	09 24' E est.
12	FECHENHEIM	8,036	16-Mar-46	58 14' N est.	09 24' E est.
13	FREIBURG	6,500	13-Jul-46	Unk	Unk
14	GEORGE HAWLEY	1,120	30-Jun-47	58 18' 05" N	09 38' 00" E
15	GERTRUD FRITZEN	4,500	13-Jul-46	Unk	Unk
16	HUGO OLDENDORF	8,037	16-Mar-46	58 14' N est.	09 24' E est.
17	JAMES HARROD	3,360	20-Jun-47	58 16' 00" N	09 33' 00" E
18	JAMES OTIS	4,091	30-Aug-46	58 16' 00" N	09 32' 00" E
19	JAMES SEWELL	4,480	06-Jun-47	58 15' 02" N	09 30' 06" E
20	JANTJE FRITZEN	6,600	17-Nov-45	58 15' N	09 30' E
21	KARL LEONHARDT	8,000	16-Mar-46	58 14' N	09 24' E
22	LOUISE SCHROEDER	1,800	04-Oct-45	58 15' N	09 27' E
23	LUDWIGSHAVEN	1,721	12-Oct-46	58 17' N	09 38' E
24	MONTE PASCOAL	6,000	21-Dec-46	58 10.31 N	10 46.13 E
25	NESBITT	6,720	18-Jul-47	58 15' 00" N	09 30' 00" E
26	ODERSTROM	2,465	17-Oct-45	58 16' N est.	09 26' E est.
27	OLGA SIEMERS	5,000	17-Oct-45	58 16' N est.	09 26' E est.
28	PATAGONIA	8,000	04-Oct-45	58 15' N	09 35' E
29	PILLAU	1,800	04-Oct-45	58 15' N est.	09 30' E est.
30	RHON	1,061	08-Sep-46	58 17' N	09 36' 45" E
31	SCHWABENLAND	6,000	06-Jun-47	58 10.22 N	10 45.24 E
32	SESOSTRIS	2,000	17-Nov-45	58 18.315 N	09 41.057 E
33	SPERRBRECHER	1,511	01-Jul-46	58 14' 00" N	09 15' 00" E
34	T-63	6,000	21-Dec-46	Unk	Unk
35	T-65	1,709	01-Jul-46	58 17' 09" N	09 37' 01" E
36	TAGILA	2,600	17-Nov-45	58 15' N	09 30' E
37	TAURUS	1,000	17-Nov-45	58 16.009 N	09 31.152 E
38	THEDA FRITZEN	2,466	17-Nov-45	58 18' N	09 55' E
39	TRITON	2,000	04-Oct-45	58 15' N est.	09 30' E est.
40	TRUDE SCHUNEMANN	1,500	17-Oct-45	58 16' N est.	09 26' E est.
41	U-J 305	752	02-Jul-46	58 16' 04" N	09 29' 00" E

Figure 105
Alphabetical Listing of the 41 CW Vessels
Scuttled by the U.K. and U.S. in the Skagerrak and Kattegat

NORWEGIAN SEA					
			Date		
		Tonnage	Scuttled	Latitude	Longitude
1	MARCY	2,800	24-Aug-48	62 59' 00" N	01 23' 00" E
2	PHILIP HEINIKEN	2,240	24-Jul-48	62 57' 00" N	01 32' 00" E

Figure 106
Listing of the 2 CW Vessels
Scuttled by the U.S. in the Norwegian Sea

ATLANTIC OCEAN (N. of Ireland and Scotland)					
			Date		
		Tonnage	Scuttled	Latitude	Longitude
1	BOTLEA / AFRICAN PRINCE		30-Dec-45	55 30.00 N	11 00.00 W
2	EMPIRE CLAIRE		27-Jul-55	56 30.00 N	12 00.00 W
3	EMPIRE CORMORANT		1-Oct-45	55 30.00 N	11 00.00 W
4	EMPIRE FAL		2-Jul-45	55 00.09 N	11 00.00 W
5	EMPIRE SIMBA		11-Sep-45	55 30.00 N	11 00.00 W
6	EMPIRE WOODLARK		2-Nov-46	59 00.00 N	07 40.00 W
7	KOTKA		23-Jul-56	56 31.00 N	12 05.00 W
8	LEIGHTON		09-Aug-47	56 22.00 N	09 27.00 W
9	VOGTLAND		30-May-56	56 30.00 N	12 00.00 W
10	WAIRUNA		30-Oct-45	55 30.00 N	11 00.00 W
11	Unknown Ship Name #1		1956	56 00.00 N	10 00.00 W
12	Unknown Ship Name #2		1956	56 00.00 N	10 00.00 W

Figure 107
Listing of the 12 CW Vessels
Scuttled by the U.K. in the Atlantic Ocean

BAY OF BISCAY					
			Date		
		Tonnage	Scuttled	Latitude	Longitude
1	DORA OLDENDORFF		05-Feb-47	47 40.00 N	09 22.00 W
2	EMPIRE CONNYNGHAM		20-Jun-49	47 52.00 N	08 51.00 W
3	EMPIRE LARK		22-Jul-47	47 55.00 N	08 17.00 W
4	EMPIRE NUTFIELD		3-Sep-46	48 03.00 N	08 09.00 W
5	EMPIRE PEACOCK		25-Aug-46	47 57.00 N	08 33.24 W
6	EMPIRE SUCCESS		22-Aug-48	47 16.30 N	09 24.00 W
7	HARM FREITZEN		01-Mar-48	47 55.00 N	08 58.00 W
8	KINDERSLEY		01-Oct-46	47 54.00 N	08 21.00 W
9	LANARK		11-Nov-46	48 00.00 N	08 21.00 W
10	MARGO		03-Nov-47	47 36.00 N	09 31.00 W
11	MIERVALDIS		22-Sep-48	47 23.00 N	09 24.00 W
12	THORPE BAY		08-Sep-47	47 47.30 N	08 21.00 W

Figure 108
Listing of the 12 CW Vessels Scuttled by the U.K. in the Bay of
Biscay

SUSPECTED IN SKAGERRAK / KATTEGAT			Date		
		Tonnage	Scuttled	Latitude	Longitude
1	BERLIN		31-May-46	57 08' 50" N	10 49' 12" E
2	BERNLEF / WAR OLIVE		14-Aug-45	56 10' 00" N	12 07' 00" E
3	BREMSE		Unk	57 52' 00" N	06 15' 00" E
4	CLAUS VON BEVERN (T-190)		16-Mar-46	57 52' 00" N	06 15' 00" E
5	F-192		Unk	58 08' 00" N	10 52' 00" E
6	H.C. HORN		26-May-46	58 08' 30" N	10 50' 00" E
7	JAN WELLENS		Oct-45	58 00' 00" N	09 30' 00" E
8	KRYSSER LEIPZIG	1,000	16-Dec-46	57 52.011 N	06 15.747 E
9	LOTTE		26-Mar-46	58 19' 00" N	09 40' 00" E
10	M-16		18-May-46	58 10' 12" N	10 42' 24" E
11	M-280		26-Jul-46	57 40' 00" N	06 30' 00" E
12	M-522		18-May-46	58 10' 12" N	10 40' 48" E
13	S-12		02-May-46	58 09' 00" N	10 52' 00" E
14	S-7		02-May-46	58 09' 00" N	10 50' 00" E
15	S-9		02-May-46	58 09' 00" N	10 51' 00" E
16	T-21		16-Dec-46	57 52' 00" N	06 15' 00" E
17	T-37		26-Jul-46	57 40' 00" N	06 30' 00" E
18	T-38		10-May-46	58 07' 48" N	10 46' 30" E
19	T-39		10-May-46	58 08' 12" N	10 47' 48" E
20	TF-1		02-May-46	58 09' 15" N	10 50' 30" E
21	Z-29		16-Dec-46	57 52' 00" N	06 15' 00" E
22	Z-34		26-Mar-46	58 19' N est.	09 40' E est.

Figure 109
Listing of the 22 Additional Vessels
Suspected to Have Been Scuttled in the Skagerrak and
Kattegat with CW Cargoes

OTHER SUSPECTED			Date		
		Tonnage	Scuttled	Latitude	Longitude
1	ARTHUR SEWALL		12-Oct-46	Unk	Unk
2	BRANDENBURG (VS-158)		Unk	Unk	Unk
3	ERIKA SCHUNEMANN		Unk	Unk	Unk
4	GEMLOCK		Unk	Unk	Unk
5	HELGOLAND		1948	Unk	Unk
6	HERBERT NORKUS		1947	Unk	Unk
7	JAMES W. NESMITH		03-Jun-46	Unk	Unk
8	KSB-1		Unk	Unk	Unk
9	KSB-13		Unk	Unk	Unk
10	OCEAN TRANSPORT 2		Unk	Unk	Unk
11	T-156		03-May-45	Unk	Unk

Figure 110
Listing of the 11 Additional Ships Which May Have Been
Included in CW Disposal Operations

CHAPTER 5

ESTIMATED TOTAL CHEMICAL WARFARE AGENTS DISPOSED IN EUROPEAN WATERS

"There is an ever-growing awareness among all sections of society about the carelessness in handling chemicals which has precipitated a precarious environmental situation. In coming decades this may lead to a general ecological crisis and pose a serious threat to the continued existence of the human race – perhaps the gravest in its history. CW agents and other militarily important poisons are today capable of bringing about the self-annihilation of man via their acute effects and, particularly, via their delayed effects on human health and the environment."

"Those working in stores where residual stockpiles of such CW agents are still being kept (organoarsenic agents such as Clark I, II and Adamsite) – or at places where sunken or buried deposits from World War I and II someday come to light – run long-term health risks; not to mention the ecological consequences of arsenical poisoning of the environment. The great stability of arsenic compounds in soil and water points to the continued imminence of an arsenical hazard to health even in coming decades."

PROFESSOR DR. KARLHEINZ LOHS
Delayed Toxic Effects of Chemical Warfare Agents
SIPRI, 1975

"The sea knows no boundaries – and
neither does pollution."
Danish Environmental Protection Agency
International Agreements on the Marine Environment

The most widely accepted definition of marine pollution is the one devised by the United Nations Joint Group of Experts on the Scientific Aspects of Marine Pollution (GES-AMP). This states that marine pollution is the "Introduction by man, directly or indirectly, of substances or energy into the marine environment (including estuaries) resulting in such deleterious effects as harm to living organisms, hazards to human health, hindrance to marine activities including fishing, impairment of quality for use of sea water and reduction of amenities."

"I concur in your concern about dumping deadly nerve gas in the ocean and urge you marshal support for further investigation and for consultations with marine biologists and scientists about the ecological hazards involved. It seems that little minded men with intellectual and philosophical blind spots have a compulsive need to make a garbage dump out of the ocean for everything man doesn't want. How can we know that the deadly nerve gas will not have disastrous effects on marine life when metal encasements of the bombs corrode away 25 years from now? Let's try to stop the ecological madness and stupidity of our present generation of men who would recklessly pollute the air, soil, and the oceans. Let's think things through in depth before taking action. Urge you send copy of this message to President Nixon." [129]

DR. H. WELDON WILKINSON
1969

5.1 Estimates of National Authorities and International Organizations

According to the Report to the 15th Meeting of the Helsinki Commission from the Ad-Hoc Working Group on Dumped Chemical Munition (HELCOM CHEMU), it is not possible to specify an accurate net weight of the warfare agents alone because of the following factors:

- The quantity of chemical warfare agents varies for the individual types of munitions, depending on their purpose. The decisive factor is whether the munitions are artillery ammunition, aerial bombs, or other containers which consist only of a thin shell and thus contain a larger amount of warfare agents;
- Some of the objects were intact munitions filled with chemical warfare agents and some consisted of empty shells with no chemical warfare agents, and
- Information about the composition of the various munitions cargoes is highly inadequate.

In the Introduction, the critical question "What amount of toxic chemical agents remain?" was posed. Here are the findings and estimates of a number of national authorities and international organizations:

In May 1993, the German Federal Maritime and Hydrographic Agency stated in their report "Chemical Munitions in the Southern and Western Baltic Sea": "The amount of warfare agent in relation to the total weight of the various types of munitions ranged between 3% and

60%; the average for the chemical munitions as a whole is around 15%.

The October 1993 National Report of the Russian Federation, "Complex Analysis of the Hazard Related to the Captured German Chemical Weapon (sic) Dumped in the Baltic Sea", in addition to reporting the gross tonnage of German CW captured in the Russian Zone, lists in precise detail the total amounts of toxic chemical agents dumped into the Baltic Sea in 1947 – by munition type and agent. Official records indicate a total of 12,035 tons of chemical agents were sea disposed in 1947.[130] The Table at Figure 111 provides subtotals of chemical agents sea disposed in each principal region as identified in section 3.2.3.[131]

Based upon these accurate records (12,035 tons of chemical agents and 70,500 gross tons), chemical agents accounted for 17.1% of the gross weight of the munitions (12,035/70,500).[132]

TYPE OF CW	AGENTS					
Region and Subtotals	H	Arsinöl	DM	CN	Other	TOTAL
Aircraft Bombs						
1	512	78	51	41		682
2	5,920	906	591	479		7,896
	6,432	984	642	520	0	8,578
Artillery Shells						
1	58	0	5		3	66
2	671	0	61	36	0	768
	729	0	66	36	3	834
HE Bombs						
1	27	0	0	0	0	27
2	314	0	0	0	0	314
	341	0	0	0	0	341
Mines						
1	4	0	0	0	0	4
2	42	0	0	0	0	42
	46	0	0	0	0	46
Barrels						
1	7	18	60	0	6	91
2	80	203	693	0	74	1,050
	87	221	753	0	80	1,141
Smoke Grenades						
1	0	0	6	0	0	6
2	0	0	65	0	0	65
	0	0	71	0	0	71
Containers						
1	0	80	0	0	0	80
2	0	924	0	0	0	924
	0	1,004	0	0	0	1,004
Drums						
1	0	0	2	0	0	2
2	0	0	18			18
	0	0	20	0	0	20
TOTAL Region 1:	608	176	124	41	9	958
TOTAL Region 2:	7,027	2,033	1,428	515	74	11,077
GRAND TOTAL:	7,635	2,209	1,552	556	83	12,035

Figure 111
Total Tonnage of Chemical Warfare Agents
Disposed in the Baltic Sea, 1947

Despite citing reasons why estimating an average percentage for chemical agents is difficult, the Ad Hoc Working Group on Dumped Chemical Munition (sic) did make the following assessment in their January 1994 Report to the 15th Meeting of the Helsinki Commission: "Several countries within the Helsinki Convention estimate the content of active chemical warfare agents in artillery rounds to be 10% and 60% for bombs. Consequently, it is assumed that 10-20% of the known quantities of chemical munitions dumped are active chemical warfare agents."[133]

In the study "The Challenge of Old Chemical Munitions and Toxic Armament Wastes" published by the Stockholm International Peace Research Institute (SIPRI) in 1997, it was estimated that chemical agents account for 10-25% of the gross weight of the munitions.[134]

Considering these estimates and ranges (10-25%), if approximately 754,975 tons (gross weight) of chemical weapons were disposed in European waters after WWII, it can therefore be estimated that between 75,497 and 188,744 tons of chemical agents (net weight) were dumped.

Given that the Russian Baltic Disposal Operations in 1947 (70,500 tons) alone account for 9.3% of the total amount of CW estimated to have been disposed in European waters post WWII, if their precise calculation of 17.1% chemical agent weight is used, 129,100 tons of chemical agents (net weight) were dumped.

5.2 Potential Environmental Issues

Sea-disposed munitions pose at least three basic types of danger:

1. Direct physical contact with either chemical or conventional munitions resulting in threats to human health. Direct physical contact can occur in a number of ways and includes individuals who are involved in working in close proximity to dumped munitions such as fishermen, pipeline layers, and those involved in construction projects such as dredging or off-shore wind farms. Of particular concern are encounters by fishermen with mustard gas, oftentimes with reportedly severe results.

2. Contamination of marine organisms and the environment in the vicinity of dumped munitions and the consequent potential for concentrations of toxic contaminants entering the wildlife and human food chains.

3. Spontaneous explosions which can be directly life threatening and also have the potential of spreading the material away from the dump sites thereby increasing the potential for more of the toxic chemical agents to come into direct physical contact with individuals.[135]

It is important to note here as well that in aquatic systems, the transport, behavior, fate and therefore exposure of biota to chemical contaminants are governed by the physical and chemical properties of each of the agents involved as well as their unpredictable interactions.

Also important to note is the fact that "The marine ecosystem is not comparable with the laboratory environment, and little is still known about the dynamic behavior of pollutants under actual marine conditions, their environmental impact and possible bioaccumulation in fauna and flora."[136]

Because of the magnitude of information now available regarding potential environmental impact – and since the sea disposal of chemical weapons is a critical global issue, this protocol will be addressed in detail in the second volume of this research: "Global Disposal Operations and Environmental Impact".

CHAPTER 6

LEGAL RESPONSIBILITIES OF STATES

6.1 The Evolution of International Policy, Protocols, and Laws

Disposal of chemical weapons has occurred in every ocean. It was not until the Convention on the High Seas was promulgated in 1958, however, that the general issue of sea dumping was dealt with in the international community.

In the 1970s several additional agreements were concluded, reflecting the growing international awareness of damage done to the environment, including ocean pollution. In the United States, the Marine Protection, Research, and Sanctuaries Act of 1972 (Public Law 92-532) outlawed U.S. dumping of chemical weapons. Nonetheless, dumping of chemical weapons was not explicitly addressed internationally until negotiations towards a Chemical Weapons Convention (CWC) occurred in the 1980s. In the current CWC, there is a provision that specifically prohibits sea disposal of chemical weapons as a means of destroying chemical weapons stockpiles. Descriptions of pertinent international agreements addressing the sea disposal of hazardous substances follow:

Link to the International Maritime Organization (IMO) "Status of Conventions" Chart:

http://www.imo.org/About/Conventions/StatusOfConventions/Pages/Default.aspx

Convention on the High Seas
29 April 1958
http://untreaty.un.org/ilc/texts/instruments/english/ conventions/8_1_1958_high_seas.pdf

The Convention on the High Seas was designed to codify rules of international law relating to the high seas. The Convention was adopted by the United Nations Conference on the Law of the Sea (UNCLOS I) which was held February 24 - April 27, 1958. The treaty was signed 29 April 1958 and entered into force 30 September 1962. With respect to sea dumping, Article 2 defines a standard of "reasonable regard" by which signatories should abide.

"These freedoms, and others which are recognized by the general principles of international law, shall be exercised by all States with reasonable regard to the interests of other States in their exercise of the freedom of the high seas."

Paragraph 2 of Article 25 of the Convention specifically addresses the requirement for States to "cooperate with the competent international organizations in taking measures for the prevention of pollution of the seas or air space above, resulting from any activities with radioactive materials or other harmful agents."

Of the principal dumping States, the United Kingdom signed on 9 September 1958, the United States on 15 September 1958 and both Germany and the Russian signed on 30 October 1958.

Link to the List of Signatories to the Convention on the High Seas:
http://treaties.un.org/doc/publication/mtdsg/volume %20ii/chapter%20xxi/xxi-2.en.pdf

US/USSR Executive Bilateral Agreement
Relating to Fisheries in the Northeastern Part
of the Pacific Ocean off the U.S. Coast
13 February 1967
http://iea.uoregon.edu/pages/view_treaty.php?t=1967-
USUSSRCertainFisheryProblemsNortheasternPacificOf
fUSCoast.EN.txt&par=view_treaty_html

Paragraph 7 of this agreement reads "Both Governments will take appropriate measures to ensure that, to the extent practicable, waste materials are discharged at sea only in water deeper than 1000 meters." As of the May 14, 1969 Hearing on the International Implications of Dumping Poisonous Gas and Waste into Oceans before the Subcommittee on International Organizations and Movements, Committee on Foreign Affairs, U.S. House of Representatives, this was the only anti-contamination clause in eight international fisheries conventions and seven bilateral agreements. This clause is intended to control the dumping of waste materials by large Soviet fishing vessels in ground fished by U.S. fishermen. The Soviet practice had made certain grounds less attractive for U.S. fishermen to fish due to the danger of fouling up their gear and the waste materials brought up in their catches.

Hearing on the International Implications of Dumping Poisonous Gas and Waste into Oceans
Subcommittee on International Organizations and Movements
Committee on Foreign Affairs, U.S. House of Representatives
May 8, 13, 14, and 15, 1969

On May 14, 1969 of this Hearing, then Acting Deputy Legal Advisor for the U.S. Department of State, Mr. Richard Frank, was asked if the U.S. Department of Defense (DoD) planned to dump chemical weapons off the coast of New Jersey in violation of the Convention of the High Seas. In his response he suggested "...Article 25 is relevant in that it requires cooperation with international organizations. But...Article 2 requires reasonableness in what we're doing. If it is unreasonable, then regardless of the cooperation, we can't do it. There would be a rule of international law that would inhibit our doing it." It was later determined the CW dumping that occurred in 1970 was "reasonable". In addition, the 1993 U.S. Arms Control and Disarmament Agency (ACDA) report claims the U.S. Army observed a 1969 National Academy of Sciences (NAS) recommendation to cease dumping at sea after the final CHASE dump took place. Details of the 1970 disposal:

18 August 1970
- 12,508 M55 rockets containing sarin (GB)
- 3 155mm projectiles containing sarin (GB)
- 1 M23 land mine containing nerve agent VX

Sunk in the hulk of S.S. LeBaron Russell Briggs. Rockets were packaged in steel-enclosed concrete vaults and loaded aboard the hulk at the Military Ocean Terminal, Sunny Point, South Carolina. Disposal Site: Atlantic Ocean: 29°23.1'N, 75°58'W at a depth of 16,400 ft. (5,000 m)

It is noted that chemical weapons were dumped in the sea in Operation CHASE (Cut Holes And Sink 'Em) three times prior to 1970, without obtaining the advice of the State Department with respect to international law.

15 June 1967

- 4,577 Ton containers of mustard (H)
- 7,380 M55 rockets of sarin (GB) in concrete blocks

Sunk in the hulk of S.S. Corporal Eric G. Gibson
Disposal Site: Atlantic Ocean: 39°38.7'N, 70°57'W

19 June 1968

- 38 Ton containers of sarin (GB) and nerve agent (VX)
- 1,460 vaults of M55 rockets of Sarin (GB) and nerve agent (VX)
- 120 drums and canisters of arsenic and cyanide

Sunk in the hulk of S.S. Mormactern
Disposal Site; Atlantic Ocean: 39°32.7'N, 71°01.6'W at a depth of 6,390 ft. (1,948 m)

<u>7 August 1968</u>

- 3,500 Ton containers of mustard (H)

Sunk in the hulk of S.S. Richardson
Disposal Site: Atlantic Ocean: 29°23.1N, 75°58'W
at a depth of 7,800 ft. (2,377 m)

National Environmental Policy Act
U.S. Congress
(Public Law 91-190)
1 January 1970
http://www.usinfo.org/enus/government/branches/nepaeqi
a.htm

The National Environmental Policy Act (NEPA) was signed into law on January 1, 1970 by then President Nixon. The Act requires preparation of Environmental Impact Statements (EIS) for major federal actions significantly affecting the environment. It also created the Environmental Protection Agency (EPA) and the Council on Environmental Quality (CEQ).

Sec. 102 [42 USC § 4332] The Congress authorizes and directs that, to the fullest extent possible:

(1) the policies, regulations, and public laws of the United States shall be interpreted and administered in accordance with the policies set forth in this Act, and

(2) all agencies of the Federal Government shall --

> (A) utilize a systematic, interdisciplinary approach which will insure the integrated use of the natural and social sciences and the environmental design arts in planning and in decision-making which may have an impact on man's environment;

> (B) identify and develop methods and procedures, in consultation with the Council on Environmental Quality established by title II of this Act, which will insure that presently unquantified environmental amenities and values may be given appropriate consideration in decision-making along with economic and technical considerations;

> (C) include in every recommendation or report on proposals for legislation and other major Federal

actions significantly affecting the quality of the human environment, a detailed statement by the responsible official on --

(i) the environmental impact of the proposed action,

(ii) any adverse environmental effects which cannot be avoided should the proposal be implemented,

(iii) alternatives to the proposed action,

(iv) the relationship between local short-term uses of man's environment and the maintenance and enhancement of long-term productivity, and

(v) any irreversible and irretrievable commitments of resources which would be involved in the proposed action should it be implemented.

Armed Forces Authorization Act, 1970
U.S. Congress
(Public Law 91-441)
http://uscode.house.gov/download/pls/50C32.txt

Public Law 91-441 prohibits the disposal of chemical and biological agents within or outside of the United States unless they have been detoxified or made harmless to man and the environment, except in an emergency to safeguard human life. An immediate report should be made to Congress in the event of such a disposal. This law also prohibits disposal of any munitions in international waters.

50 USC Sec. 1518

TITLE 50 - WAR AND NATIONAL DEFENSE

CHAPTER 32 - CHEMICAL AND BIOLOGICAL WARFARE PROGRAM

STATUTE:

On and after October 7, 1970, no chemical or biological warfare agent shall be disposed of within or outside the United States unless such agent has been detoxified or made harmless to man and his environment unless immediate disposal is clearly necessary, in an emergency, to safeguard human life. An immediate report should be made to Congress in the event of such disposal.

SOURCE:

Public Law 91-441, Title V, Sec. 506(d), Oct. 7, 1970, 84 Stat. 913

Seabed Arms Control Treaty
Treaty on the Prohibition of the Emplacement of Nuclear Weapons and Other Weapons of Mass Destruction on the Seabed and the Ocean Floor and in the Subsoil Thereof
11 February 1971
http://www.state.gov/www/global/arms/treaties/seabed1.html

The Seabed Arms Control Treaty was intended to prevent an arms race on the ocean floor and to preserve the seabed for peaceful purposes. The basic obligation contained in paragraph 1 of Article 1 is as follows:

"The States Parties to this Treaty undertake not to implant or emplace on the seabed and the ocean floor and in the subsoil thereof beyond the outer limit of a seabed zone, as defined in Article II, any nuclear weapons or any other types of weapons of mass destruction as well as structures, launching installations, or any other facilities specifically designed for storing, testing, or using such weapons."

The U.S. Department of State determined this Treaty to be inapplicable to the sea disposal of chemical weapons, arguing that while dumping of CW could be viewed as "emplacement", dumping was never intended to enable the use of chemical weapons, but rather to render them militarily useless.

Of the principal dumping States, the United Kingdom, United States, and the U.S.S.R. signed on 11 February 1971. The Federal Republic of Germany signed on 8 June 1971.

Link to the List of Signatories to the Seabed Arms Control Treaty:

http://www.state.gov/www/global/arms/treaties/seab ed3.txt

Declaration on the Human Environment
The Stockholm Conference on the Human Environment
5-16 June 1972
**http://www.unep.org/Documents.Multilingual/Defa
ult.asp?documentid=97&articleid=1503**

The Stockholm Conference on the Human Environment recommended the establishment of the U.N. Environment Program (UNEP) and also issued a Declaration of the Human Environment enumerating 26 principles (of which the first 25 were approved by acclamation and the 26th by a separate vote, against opposition by China).

Principle 7 reads: "States shall take all possible steps to prevent pollution of the seas by substances that are liable to create hazards to human health, to harm living resources and marine life, to damage amenities or to interfere with other legitimate uses of the sea."

The U.S. Arms Control and Disarmament Agency (ACDA), U.S. Department of State, in its January 19, 1993 report "Special Study on the Sea Disposal of Chemical Munitions", noted that "although there was no binding agreement resulting from the conference, 26 principles were acclaimed."

Link to the 26 February 1971 "Report of the Preparatory Committee for the UN Conference on the Human Environment":
http://www.unlibrary-nairobi.org/PDFs/A-CONF48PC9.pdf

Marine Protection, Research, and Sanctuaries Act of 1972
U.S. Congress
(Public Law 92-532)
23 October 1972
http://epw.senate.gov/mprsa72.pdf
http://sanctuaries.noaa.gov/management/pdfs/pl92_532.pdf

The U.S. Marine Protection, Research, and Sanctuaries Act, which was signed into law by the President in 1972 (Public Law 92-532), clearly prohibits the sea disposal of any chemical agents from the United States and similarly prohibits any cooperation in the sea disposal of chemical agents from outside the United States.

"No person shall transport from the United States any radiological, chemical, or biological warfare agent...for the purpose of dumping into ocean waters." Further, "No officer, employee, agent, department, agency, or instrumentality of the United States shall transport from any location outside the United States any radiological, chemical, or biological warfare agent...for the purpose of dumping it into ocean waters. (Section 101, 33 U.S. Code 1411)

(a) Dangers of unregulated dumping: Unregulated dumping of material into ocean waters endangers human health, welfare, and amenities, and the marine environment, ecological systems, and economic potentialities.

(b) Policy of regulation and prevention or limitation: The Congress declares that it is the policy of the United States to regulate the dumping of all types of materials into ocean waters and to prevent or strictly limit the dumping into ocean waters of any material which would adversely affect human health, welfare, or amenities, or the marine environment, ecological systems, or economic potentialities.

(c) Regulation of dumping and transportation for dumping purposes: It is the purpose of this Act to regulate (1) the transportation by any person of material from the United States and, in the case of United States vessels, aircraft, or agencies, the transportation of material from a location outside the United States, when in either case the transportation is for the purpose of dumping the material into ocean waters, and (2) the dumping of material transported by any person from a location outside the United States, if the dumping occurs in the territorial sea or the contiguous zone of the United States.

Link to the Congressional Research Service (CRS) Report: "Ocean Dumping Act: A Summary of the Law", 15 December 2010:
http://assets.opencrs.com/rpts/RS20028_20101215.pdf

Convention on the Prevention of Marine Pollution by Dumping of Wastes and Other Matter "The London Convention"
"LC 72"

Opened for Signature: 29 December 1972
Entered into Force: 30 August 1975
http://www.iaea.org/Publications/Documents/Infcircs/Oth
ers/inf205.shtml
**http://www.imo.org/About/Conventions/ListOfConv
entions/Pages/Convention-on-the-Prevention-of-
Marine-Pollution-by-Dumping-of-Wastes-and-
Other-Matter.aspx**

The London Convention of 1972 defines dumping as the deliberate disposal of waste from ships and aircraft, but excluding the disposal of waste from ships and aircraft. Wastes are divided into three categories.

The first category (the "black list") includes substances such as organo-halogen compounds, mercury, cadmium, oil, plastics, and high-level radioactive wastes defined by the International Atomic Energy Agency (IAEA) as unsuitable for dumping. Dumping of substances on the black list is prohibited.

The second category (the "grey list") includes substances such as arsenic, lead, copper, zinc, organosilicon compounds, cyanides, flourides, pesticides, scrap metal, and radioactive matter not included on the black list. The dumping of these substances is permitted only if a prior special permit (issued by the national authorities of a contracting party) has been obtained.

The third category includes all wastes not on the black or grey lists. Such wastes may nevertheless be dumped only if a prior general permit has been obtained.

The U.S. Arms Control and Disarmament Agency (ACDA), U.S. Department of State, in its January 19, 1993 report "Special Study on the Sea Disposal of Chemical Munitions", states "this agreement does not apply to U.S. activities since the U.S. did not conduct any sea dumping of chemical weapons after 1970.

It is noted the military use of herbicides in Vietnam began in 1961, was expanded during 1965 and 1966, and reached a peak from 1967 to 1969. Herbicides were used extensively in Vietnam by the U.S. Air Force's Operation RANCH HAND to defoliate inland hardwood forests, coastal mangrove forests, and cultivated land, by aerial spraying from C-123 cargo/transport aircraft and helicopters. The purpose of spraying herbicides was to improve the ability to detect enemy base camps and enemy forces along lines of communication and infiltration routes. Spraying was also used to destroy the crops of the Viet Cong and North Vietnamese. The code name for the overall herbicide program was TRAIL DUST. The code name RANCH HAND specifically referred to the C-123 herbicide spraying project.

The different types of herbicide used by U.S. forces in Vietnam were identified by a code name referring to the color of the 4-inch band painted around the 55-gallon drum that contained the chemical. These included Agents Orange, White, Purple, Blue, Pink, and Green. e.g. A 55-gallon drum with an orange band contained 50% n-butyl ester of 2,4-D (2,4-dichlorophenoxyacetic acid) and 50% n-butyl or isooctyl ester of 2,4,5-T(2,4,5-trichlorophenoxyacetic acid).

Agent Orange accounted for over 60% of the total herbicides disseminated over Vietnam (11.7 million gallons of a total 19.4 million gallons) and contained relatively high levels of an exceedingly poisonous contaminant known as "dioxin" or "TCDD" (2,3,7,8-tetrachlorodibenzo-p-dioxin).

At the conclusion of the Vietnam War, approximately 40,275 55-gallon drums of Agent Orange remained. The 15,480 drums of the agent stockpiled at the Naval Construction Battalion Center (NCBC) in Gulfport, Mississippi were transferred to the Dutch-owned ship, the Vulcanus, and incinerated at sea between 15-24 July 1977. The 24,795 drums of the agent stored on Johnston Island in the Pacific were subsequently incinerated on the Vulcanus in two loads. The last of the herbicide orange once destined for the jungles of Vietnam burned on September 3, 1977.

It was not until 1978, however, that the parties to the London Convention adopted amendments to the annexes dealing with the incineration of wastes at sea. A special permit is required for the incineration of wastes on both the black and grey lists and such incineration must comply with the regulations set out in the revised annexes.

Amendment to the
Marine Protection, Research, and Sanctuaries Act
of 1972
London Dumping Convention Implementation
U.S. Congress
(Public Law 93-254)
22 March 1974
http://upload.wikimedia.org/wikipedia/commons/th
umb/7/7f/United_States_Statutes_at_Large_Volume_
88_Part_1.djvu/page94-949px-
United_States_Statutes_at_Large_Volume_88_Part_
1.djvu.jpg

Public Law 93–254 struck out statement of the purpose of this Act as being the regulation of transportation of material from the United States for dumping into ocean waters, and the dumping of material, transported from outside the United States, if the dumping occurs in ocean waters over which the United States has jurisdiction or over which it may exercise control, under accepted principles of international law, in order to protect its territory or territorial sea, now covered by subsection (c) of this section. Public Law 93–254 added subsection. (c).

33 U.S. Codes, Section 1401

http://www.gpo.gov/fdsys/pkg/USCODE-2010-
title33/pdf/USCODE-2010-title33-chap27-sec1402.pdf

http://www.law.cornell.edu/uscode/text/33/1401

Congressional finding, policy, and declaration of purpose

(a) Dangers of unregulated dumping

Unregulated dumping of material into ocean waters endangers human health, welfare, and amenities, and the marine environment, ecological systems, and economic potentialities.

(b) Policy of regulation and prevention or limitation

The Congress declares that it is the policy of the United States to regulate the dumping of all types of materials into ocean waters and to prevent or strictly limit the dumping into ocean waters of any material which would adversely affect human health, welfare, or amenities, or the marine environment, ecological systems, or economic potentialities.

(c) Regulation of dumping and transportation for dumping purposes

It is the purpose of this Act to regulate (1) the transportation by any person of material from the United States and, in the case of United States vessels, aircraft, or agencies, the transportation of material from a location outside the United States, when in either case the transportation is for the purpose of dumping the material into ocean waters, and (2) the dumping of material transported by any person from a location outside the United States, if the dumping occurs in the territorial sea or the contiguous zone of the United States.

References in Text

This Act, referred to in subsec. (c), means Pub. L. 92–532, which is classified generally to this chapter, chapter 41 (§2801 et seq.) of this title, and chapters 32 (§1431 et seq.) and 32A (§1447 et seq.) of Title 16, Conservation.

Amendments

1974—Subsec. (b). Pub. L. 93–254 struck out statement of the purpose of this Act as being the regulation of transportation of material from the United States for dumping into ocean waters, and the dumping of material, transported from outside the United States, if the dumping occurs in ocean waters over which the United States has jurisdiction or over which it may exercise control, under accepted principles of international law, in order to protect its territory or territorial sea, now covered by subsec. (c) of this section.

Subsec. (c). Pub. L. 93–254 added subsec. (c).

Effective Date of 1974 Amendment

Section 2 of Pub. L. 93–254 provided in part that amendment of subsecs. (b) and (c) of this section and sections 1402, 1411, and 1412(a), other than last sentence of subsec. (a), of this title, by Pub. L. 93–254 shall become effective Mar. 22, 1974.

Convention on the Prohibition of Military or Any Other Hostile Use of Environmental Modification Techniques
(ENMOD Convention)
Opened for Signature: 18 May 1977
Entered into Force: 5 October 1978
http://disarmament.un.org/treaties/t/enmod/text

The ENMOD Convention seeks to prevent the use of environmental modification techniques such as climate control for hostile purposes. In particular, the ENMOD defines modification techniques as deliberate manipulation of natural processes. Paragraph 1 of Article 1 states that:

"Each State Party to this Convention undertakes not to engage in military or any other hostile use of environmental modification techniques having widespread, long-lasting, or severe effects as the means of destruction, damage, or injury to any other State party."

The U.S. Arms Control and Disarmament Agency (ACDA), U.S. Department of State, in its January 19, 1993 report "Special Study on the Sea Disposal of Chemical Munitions", noted that the U.S. did not engage in sea disposal of chemical weapons as a hostile act to alter the ocean environment.

Germany, Russia, the United Kingdom, and the United States all signed the Convention on the day it opened for signature, 18 May 1977.

Link to the List of Signatories:
http://disarmament.un.org/treaties/t/enmod

United Nations Convention on the Law of the Sea (UNCLOS)
"The Law of the Sea"
1983
http://www.un.org/Depts/los/index.htm

The Law of the Sea Agreement does not specifically address the sea disposal of chemical weapons, but does contain Article 210, Pollution by Dumping. While the language does not prohibit dumping, it subjects dumping to standard rules, regulations, notification, as well as permission. It also recommends that each State adopt laws that would be no less effective than international rules regarding sea dumping.

Article 210 – Pollution by Dumping
http://www.un.org/Depts/los/convention_agreement s/texts/unclos/part12.htm

1. States shall adopt laws and regulations to prevent, reduce, and control pollution of the marine environment by dumping.
2. States shall take other measures as may be necessary to prevent, reduce, and control such pollution.
3. Such laws, regulations, and measures shall ensure that dumping is not carried out without the permission of the competent authorities of States.
4. States, acting especially through competent international organizations or diplomatic conference, shall endeavor to establish global and regional rules, standards, and recommended practices and procedures to prevent, reduce, and control such pollution. Such rules,

standards, and recommended practices and procedures shall be re-examined from time to time as necessary.
5. Dumping within the territorial sea and the exclusive economic zone or onto the continental shelf shall not be carried out without the express prior approval of the coastal State, which has the right to permit, regulate, and control such dumping after due consideration of the matter with other States which by reason of their geographic situation may be adversely affected thereby.
6. National laws, regulations, and measures shall be no less effective in preventing, reducing, and controlling such pollution than the global rules and standards.

Of the principal dumping States, Germany signed on 14 October 1994, the Russian Federation signed on 12 March 1997 and the United Kingdom signed on 25 July 1997. The United States has not yet signed.

The U.S. Arms Control and Disarmament Agency (ACDA), U.S. Department of State, in its January 19, 1993 report "Special Study on the Sea Disposal of Chemical Munitions", noted that although not a signatory, the United States abides by the provisions as customary international law.

U.S. Developments

On April 24, 2004 Jeane Kirkpatrick (Reagan Administration United Nations Ambassador 1981-1985), testified against United States ratification of the treaty before the Senate Armed Services Committee, in which she argued that "Viewed from the perspective of U.S. interests and Reagan Administration principles, it was a bad bargain," and that "its ratification will diminish our capacity for self-government, including, ultimately, our capacity for self-defense."

On April 11, 2006, the 5-Member UNCLOS Annex VII Arbitral Tribunal, presided over by H.E. Judge Stephen M. Schwebel, rendered after two years of international judicial proceedings, the landmark Barbados/Trinidad and Tobago Award, which resolved the maritime boundary delimitation (in the East, Central and West sectors) to satisfaction of both Parties and committed Barbados and Trinidad and Tobago to resolve their fisheries dispute by means of concluding a new Fisheries Agreement.

On May 15, 2007, United States President George W. Bush announced that he had urged the Senate to approve the UNCLOS.

On October 31, 2007, the Senate Foreign Relations Committee voted 17-4 to send the treaty to the full U.S. Senate for a vote.

On September 20, 2007, an Arbitral Tribunal constituted under UNCLOS issued its decision on a longstanding maritime boundary dispute between Guyana and Suriname, which contained a ruling blaming both nations for violating treaty obligations.

On January 13, 2009, speaking at her Senate confirmation hearing as nominee for United States Secretary of State, Senator Hillary Clinton said that ratification of the Law of the Sea Treaty would be a priority for her.

On May 23, 2012, Secretary Clinton testified before the Senate Committee on Foreign Relations and argued for the ratification of the treaty.

Link to the U.N. Chronological List of Ratifications of, Accessions and Successions to the Convention and the related Agreements:
http://www.un.org/Depts/los/reference_files/chronol ogical_lists_of_ratifications.htm

Convention for the Protection of the Marine Environment of the North-East Atlantic
(OSPAR Convention)
22 September 1992
www.ospar.org

As a result of the growing environmental awareness the Convention for the Prevention of Marine Pollution by Dumping from Ships and Aircraft (Oslo Convention) was signed in 1972. It entered into force in 1974. In that same year the Convention for the Prevention of Marine Pollution from Land-Based Sources (Paris Convention) was signed, entering into force in 1978. In 1992 both conventions merged into the new Oslo-Paris Convention for the Protection of the Marine Environment of the Northeast Atlantic (OSPAR Convention), which entered into force in 1998.

While the OSPAR Convention forbids the dumping of toxic waste at sea, no explicit reference is made to war material.

The OSPAR Quality Status Report (QSR) 2000, however, recognized that marine dumped munitions present a hazard to the public and that OSPAR was considering a course of action for dealing with dumped munitions.

One of the first steps taken by OSPAR was to collect information on the procedures in place by Contracting Parties in relation to marine dumped chemical weapons and munitions. To this end a questionnaire was circulated to Contracting Parties, Non-Governmental Organizations and Observers to gather information on reporting of encounters, on guidance issued to fishermen and users of the sea and on existing surveillance and management practices. The results of this exercise showed that:

- Although there are systems in place in most countries to record encounters with dumped chemical weapons and munitions, such records are not centrally maintained and/or easily accessible.
- The information contained in guidelines produced by Contracting Parties for fishermen and other users of the sea varies considerably between countries. In order to ensure that sufficient information is available to fishermen who may encounter dumped chemical weapons and munitions, the preparation of draft OSPAR guidelines was considered. (See link below)
- Given the number of dumpsites recorded in the OSPAR overview of dumped weapons, there has been relatively little monitoring to date. Where monitoring has taken place, the need for further monitoring has been identified. The requirement for further site assessments and ongoing monitoring may need to be considered by Contracting Parties.
- There is no specific protocol available to assess the risks associated with activities in the vicinity of dumpsites.
- Recovery of dumped munitions is a costly and highly risky operation which could result in the release of toxic material into the environment.

In recognition of the deficiencies in available information on the locations of marine munitions and

chemical weapons dumpsites and in the recording of encounters with munitions and chemical weapons, the OSPAR Commission published 'OSPAR Framework for reporting encounters with marine dumped chemical weapons and munitions' in 2003.
Link to the OSPAR Framework for Reporting Encounters with Marine Dumped Chemical Weapons and Munitions:

http://www.ospar.org/documents/dbase/publicati ons/p00186_reporting%20encounters%20with%20ma rine%20dumped%20chemical%20weapons.pdf

The ongoing recording and reporting of encounters was seen as a possible way of identifying previously unknown or unrecorded dumpsites and also of identifying changes in the condition of dumped material or of the dumpsite following natural and anthropogenic disturbances. This information could be used to adjust the boundaries of known dumpsites or to identify new dumpsites or areas of high risk on marine charts. This activity is also consistent with the Decision No 2850/2000/EC of the European Parliament and of the Council of 20 December 2000 setting up a Community framework for cooperation in the field of accidental or deliberate marine pollution (2850/2000/EC).

The preparation of these publications heightened the awareness among Contracting Parties that a coordinated OSPAR-wide approach would have added value in increasing our understanding of the extent and impact of past dumped chemical weapons and munitions in the entire Convention Area.[137]

Convention on the Prohibition of the Development, Production, Stockpiling and Use of Chemical Weapons and on their Destruction
"Chemical Weapons Convention"
13 January 1993
http://www.opcw.org/chemical-weapons-convention

The Chemical Weapons Convention (CWC) contains two identical provisions (Article III, Declarations, paragraph 2 and Article IV, Chemical Weapons, paragraph 17) which exclude chemical weapons dumped at sea prior to 1985:

"The provisions of this Article and the relevant provisions of Part IV of the Verification Annex shall not, at the discretion of a State Party, apply to chemical weapons buried on its territory before 1 January 1977 and which remain buried, or which had been dumped at sea before 1 January 1985."

"Since the United States has not dumped chemical weapons into the sea since 1970, there are no CWC implications with respect to our past activities." Bottom Line: The CWC has not been able to solve the problems of both pre- and post-World War II dumping operations and attendant responsibilities. It is obvious the principal States have conveniently legislated their impunity.

Little surprise that Germany, the Russian Federation, the United Kingdom, and the United States all signed the Convention on the first day it opened for signature: 13 January 1993.

Link to the Current List of Signatories:
http://www.opcw.org/about-opcw/member-states

The London Protocol
Entered into Force: 24 March 2006
http://www.imo.org/blast/blastData.asp?doc_id=132
03&filename=PROTOCOL%20Amended%202006.doc

On November 17, 1996, a special meeting of the Contracting Parties to the 1972 London Convention adopted the "1996 Protocol to the Convention on the Prevention of Marine Pollution by Dumping of Wastes and Other Matter, 1972" which is to replace the 1972 Convention, subject to ratification. In line with the United Nations Conference on Environment and Development (UNCED) (Rio Summit 1992) "Agenda 21", the 1996 Protocol reflects the global trend towards precaution and prevention with the Parties agreeing to move from controlled dispersal at sea of a variety of land-generated wastes towards integrated land-based solutions for most, and controlled sea disposal of few, remaining categories of wastes or other matter.

Among the most important innovations brought by the 1996 Protocol is the codification of the "precautionary approach" and the "polluter pays principle." Reflecting these principles, the Protocol embodies a major structural revision of the Convention—the so-called "reverse list" approach. Now, instead of prohibiting the dumping of certain (listed) hazardous materials, the Parties are obligated to prohibit the dumping of any waste or other matter that is not listed in Annex 1 ("the reverse list") of the 1996 Protocol. Dumping of wastes or other matter on this reverse list requires a permit. Parties to the Protocol are further obligated to adopt measures to ensure that the issuance of permits and permit conditions for the dumping of reverse list substances comply with Annex 2 (the Waste Assessment Annex) of the Protocol

The substances on the reverse list include dredged material; sewage sludge; industrial fish processing waste; vessels and offshore platforms or other man-made structures at sea; inert, inorganic geological material; organic material of natural origin; and bulky items including iron, steel, concrete and similar materials for which the concern is physical impact, and limited to those circumstances where such wastes are generated at locations with no land-disposal alternatives. In addition, the 1996 protocol prohibits altogether the practice of incineration at sea, except for emergencies, and prohibits the exports of wastes or other matter to non-Parties for the purpose of dumping or incineration at sea.

Article 3 – General Obligations:

1. In implementing this Protocol, Contracting Parties shall apply a precautionary approach to environmental protection from dumping of wastes or other matter whereby appropriate preventative measures are taken when there is reason to believe that wastes or other matter introduced into the marine environment are likely to cause harm even when there is no conclusive evidence to prove a causal relation between inputs and their effects.

2. Taking into account the approach that the polluter should, in principle, bear the cost of pollution, each Contracting Party shall endeavor to promote practices whereby those it has authorized to engage in dumping or incineration at sea bear the cost of meeting the pollution prevention and control requirements for the authorized activities,

having due regard to the public interest.

As of 28 May 2012, Germany signed the Protocol on 11 September 1997 and ratified it on 16 October 1998. The United Kingdom signed on 22 September 1997 and ratified on 15 December 1998. The United States signed on 31 March 1998 but has not ratified/accepted the Protocol. The Russian Federation has not signed.

Link to the IMO Guide for National Implementation: http://www.imo.org/blast/blastData.asp?doc_id=13404&filename=Guide%20to%20national%20implementation%20-%20English.doc

The "polluter pays" principle is one of the central guiding principles of the OSPAR Convention and requires that the costs of pollution prevention, control and reduction measures must be borne by the polluter. The "polluter pays" principle is mainly implemented by means of command-and-control approaches but can also be applied via market-based mechanisms, e.g. for the development and introduction of environmentally sound technologies and products.

Recognized by the ministerial North Sea Conferences in 1984, the "polluter pays" principle was included in the 1992 OSPAR Convention. Internationally the polluter pays principle was introduced in the 1970s by the Organization for Economic Cooperation and Development (OECD) and reaffirmed globally in the 1992 Rio Declaration on Environment and Development.

"Bottom Line"

It appears the principal dumping States have succeeded in hiding the truth about their CW sea disposal operations under a shroud of secrecy while quietly and conveniently legislating their impunity.

CHAPTER 7

CONCLUSION

"The significant problems we face cannot be solved at the same level of thinking we were at when we created them."
ALBERT EINSTEIN

"The sea unites nations, rather than divides them. It creates a world of neighbors."
KLAUS TÖPFER
Executive Director, United Nations Environment Program

7.1 The Imperative for an International Strategy

During its sixty-fifth session which opened on 14 September 2010, the United Nations General Assembly (UNGA) adopted a vital resolution on "Cooperative measures to assess and increase awareness of environmental effects related to waste originating from chemical munitions dumped at sea".

Link to UNGA Resolution A/RES/65/149:
http://www.un.org/ga/search/view_doc.asp?symbol= A/RES/65/149
http://www.un.org/depts/dhl/resguide/r65.shtml

Accordingly, Member States and international and regional organizations are now invited to keep under observation the issue of the environmental effects related to waste originating from chemical munitions dumped at sea and to cooperate and voluntarily share relevant information on this issue. In addition, the U.N. Secretary General is invited to seek the views of Member States and relevant regional and international organizations on issues relating to the environmental effects related to waste originating from chemical munitions dumped at sea, as well as on possible modalities for international cooperation to assess and increase awareness of this issue, and to communicate such views to the General Assembly at its sixty-eighth session for further consideration.

The unfathomable amounts of toxic chemical agents that have been dumped in European waters portend an ecocatastrophe of global proportion. The environmental and public health problems facing European nations incident to the anticipated release of potentially massive amounts of slowly hydrolyzing nerve and blister agents into the marine environment are more critical and urgent than generally supposed.

It is concluded the avoidance of the imminent problem is impossible. The toxins will leak into the marine environment. Steps can be taken, however, by all States involved, to ensure early warning and disaster preparedness. Close international cooperation of all States concerned is imperative.

7.2 Proposal and the Way Ahead

Any attempt to recover sea-disposed chemical warfare munitions now is practically infeasible. Whereas the majority of the chemical weapons disposed by the British and Americans in European waters is consolidated in hulks on the sea floor, the bulk of the CW in the Baltic was disposed piecemeal over large geographic areas. After more than 60 years of corrosion, it is assumed the munitions are too fragile. Moreover, during recovery operations as long ago as 1971-72, it was discovered that the munitions can also be under internal pressure due to gas ballast from the filling process or due to decomposition products. It must therefore be assumed that munitions are under internal pressure that can cause them to burst during a recovery operation.[138]

Given the urgent and exigent nature of the problem at hand, recommend the following systems approach be applied:

Step 1: Organize for a robust effort at the North Atlantic Treaty Organization (NATO)

Reestablish the NATO Division of Scientific and Environmental Affairs.

Reestablish the position of NATO Assistant Secretary General for Scientific and Environmental Affairs which was last held by Dr. Jean-Marie Cadiou of France. The new NATO Assistant Secretary General would:

- Assume responsibility for NATO's existing Science for Peace and Security (SPS) Program. The Science for Peace and Security Program is a policy tool for enhancing cooperation and dialogue with all partners, based on civil science and innovation, to contribute to the Alliance's core goals and to address the priority areas for dialogue and cooperation identified in the new partnership policy.

- Establish the NATO-EU Joint Committee on the Sea-Disposal of Chemical Weapons to be co-chaired with the European Commission's Director-General for the Environment.
  http://ec.europa.eu/dgs/environment/index_en .htm

The Directorate-General for the Environment is one of the more than 40 Directorates-General and services that make up the European Commission. Commonly referred to as DG Environment, the objective of the Directorate-General is to protect, preserve and improve the environment for present and future generations. To achieve this it proposes policies that ensure a high level of environmental protection in the European Union and that preserve the quality of life of EU citizens.

The Joint Committee should:

a. Ensure the scope of the investigation includes CW disposal operations in all European bodies of water.

b. Ensure the governments of the United States, United Kingdom, Russia, France, and Germany respectively establish one agency to serve as a Center for all research and data and open their archives so that the nature and exact location of all the dumps can be established.

c. Coordinate with and benefit from the work of other international fora, to include the:

- **Baltic Environmental Forum**
  http://www.befgroup.net
  The Baltic Environmental Forum (BEF) was founded in 1995 by the Baltic Ministries of Environment, Germany and the European Commission as a technical assistance project aiming at strengthening the co-operation among the Baltic environmental authorities.

- **The Baltic Marine Environment Protection Commission (Helsinki Commission – HELCOM)**
  www.helcom.fi
  The Helsinki Commission works to protect the marine environment of the Baltic Sea from all sources of pollution through intergovernmental co-operation between Denmark, Estonia, the European Community, Finland, Germany, Latvia, Lithuania, Poland, Russia and Sweden. Its Baltic Sea Action Plan is a program aimed at restoring the good ecological status of the Baltic marine environment by 2021.

- **Baltic Sea 2020**
  www.balticsea2020.org
  A private, independent foundation aimed at stimulating concrete measures that improve the environmental quality of the Baltic Sea. The assets of the foundation will finance projects that are creative, innovative and improve knowledge of the Baltic Sea until 2020.

- **ChemSec – International Chemical Secretariat**
  http://www.chemsec.org
  The International Chemical Secretariat is a non-profit organization founded in 2002 by four environmental organizations. "At ChemSec, we have an ambitious focus and goal: a toxic free environment by 2020. To achieve this, we strive to reach broad acceptance in society of the key principles of Precaution, Substitution, Polluter Pays and Right to Know."

- **Coalition Clean Baltic**
  www.ccb.se
  Coalition Clean Baltic is a cooperation of non-governmental environmental organizations from the countries of the Baltic Sea Region working to protect and improve the Baltic Sea environment and its natural resources.

- **Norwegian Defense Research Establishment (NDRE) (Forsvarets Forskningsinstitutt - FFI)**
  www.ffi.no/en
  The Norwegian Defense Research Establishment is the prime institution responsible for defense-related research in Norway. The Establishment is the chief adviser on defense-related science and technology to the Ministry of Defense and the Norwegian Armed Forces' military organization.

- **Organization for the Prevention of Chemical Weapons**
  www.opcw.org
  The Organization for the Prohibition of Chemical Weapons (OPCW) is the implementing body of the Chemical Weapons Convention (CWC). The CWC aims to eliminate an entire category of weapons of mass destruction by prohibiting the development, production, acquisition, stockpiling, retention, transfer or use of chemical weapons by States Parties. States Parties, in turn, must take the steps necessary to enforce that prohibition in respect of persons (natural or legal) within their jurisdiction.

- **OSPAR Commission**
  http://www.ospar.org
  The OSPAR Convention is the current legal instrument guiding international cooperation on the protection of the marine environment of the North-East Atlantic. It started in 1972 with the Oslo Convention against dumping. Work under the Convention is managed by the OSPAR Commission, made up of representatives of the Governments of 15 Contracting Parties and the European Commission, representing the European Union. The fifteen Governments are Belgium, Denmark, Finland, France, Germany, Iceland, Ireland, Luxembourg, The Netherlands, Norway, Portugal, Spain, Sweden, Switzerland and the United Kingdom.

- **Seas at Risk**
  www.seas-at-risk.org
  Seas at Risk is the European association of non-governmental environmental organizations working to protect and restore to health the marine environment of the European seas and the wider North East Atlantic.

- **Stockholm International Peace Research Institute (SIPRI)**
  http://www.sipri.org
  SIPRI is an independent international institute dedicated to research into conflict, armaments, arms control and disarmament. Established in 1966, SIPRI provides data, analysis and recommendations, based on open sources, to policymakers, researchers, media and the interested public.

- **United Nations Environment Program (UNEP)**
  www.unep.org
  The UNEP mission is to provide leadership and encourage partnership in caring for the environment by inspiring, informing, and enabling nations and peoples to improve their quality of life without compromising that of future generations.

With care of avoiding the duplication of assets, resources and capabilities already present within the NATO command structure, simultaneously establish the NATO Centre of Excellence (COE) for Sea-Disposed Chemical Weapons.

A COE already exists for Explosive Ordinance Disposal (EOD). Based in Trenčín, Slovakia, the Explosive Ordinance Disposal (EOD) COE provides expertise in the field of explosive ordinance disposal for NATO and Partnership for Peace countries. Like other COEs, EOD works with NATO in the areas of standardization, doctrine development and concepts validation. It supports NATO operations in the field of explosive ordinance disposal by improving interoperability and cooperation between NATO member countries, partner countries, international organizations and the NATO command structure. Established by Slovakia, the Framework Nation, in 2007, the EOD COE received NATO accreditation 28 April 2011.

Suggest a Baltic nation such as Lithuania, Latvia, or Estonia be considered the Framework nation.

Background on COEs:

Centres of Excellence (COEs) are nationally or multi-nationally funded institutions that train and educate leaders and specialists from NATO member and partner countries, assist in doctrine development, identify lessons learned, improve interoperability, and capabilities and test and validate concepts through experimentation. They offer recognized expertise and experience that is of benefit to the Alliance and support the transformation of NATO, while avoiding the duplication of assets, resources and capabilities already present within the NATO command structure.

Coordinated by Allied Command Transformation (ACT) in Norfolk, Virginia in the United States, COEs are considered to be international military organizations. Although not part of the NATO command structure, they are part of a wider framework supporting NATO Command Arrangements. Designed to complement the Alliance's current resources, COEs cover a wide variety of areas, with each one focusing on a specific field of expertise to enhance NATO capabilities.

ACT has overall responsibility for COEs and is in charge of the establishment, accreditation, preparation of candidates for approval, and periodic assessments of the centres. The establishment of a COE is a straightforward procedure. Normally, one or more members decide to establish a COE. The idea then moves into the concept development phase. During this phase the "Framework Nation" or "Nations" fleshes out the concept to ACT by providing information such as the area of specialization, the location of the potential COE and how it will support NATO transformation.

Once ACT approves the concept, the COE and any NATO country that wishes to participate in the COE's activities then negotiate two Memorandums of Understanding (MOUs): a Functional MOU, which governs the relationship between Centres of Excellence and the Alliance and an Operational MOU, which governs the relationship between participating countries and the COE. Once participating countries agree to and sign the MOU, the COE seeks accreditation from ACT.

The Alliance does not fund COEs. Instead, they receive national or multinational support, with "Framework Nations", "Sponsoring Nations" and

"Contributing Nations" financing the operating costs of the institutions. Nineteen COEs have either received NATO accreditation or are in the development stages.

Step 2: Identify and physically locate every vessel with CW cargo scuttled in European waters and every location where loose-dumping of CW occurred.

Step 3: Evaluate the dumping sites identified with regard to their environmental condition in order to accurately assess present and potential future hazardous effects on humans, animals and plants, and commercial activities.

Step 4: Formulate a strategy based upon the threat assessment.

Step 5: Install permanent monitoring stations/sensors at each dump site enabled by satellite technology – for instantaneous early warning capability.

To investigate the scuttled ships, capitalize on capabilities pioneered and refined by the Norwegian Defense Research Establishment (NDRE) in using a remote-operated vehicle (ROV) with video cameras.

To monitor all dump sites, capitalize on Copernicus – The European Earth Observation Program (previously known as GMES - Global Monitoring for Environment and Security), **http://copernicus.eu**

Copernicus, the European program for earth observation, consists of a complex set of systems which collect data from multiple sources: earth observation satellites and in situ sensors such as ground stations, airborne and sea-borne sensors. It processes these data and provides users with reliable and up-to-date information through a set of services related to environmental and security issues.

The services address six thematic areas: land, marine, atmosphere, climate change, emergency management and security. They support a wide range of applications, including environment protection, management of urban areas, regional and local planning, agriculture, forestry, fisheries, health, transport, climate change, sustainable development, civil protection and tourism.

The main users of Copernicus services are policymakers and public authorities who need the information to develop environmental legislation and policies or to take critical decisions in the event of an emergency, such as a natural disaster or a humanitarian crisis.

Based on the Copernicus services, many other value-added services can be tailored to more specific public or commercial needs. This will create new business opportunities. In fact, several economic studies so far have demonstrated a huge potential for job creation, innovation and growth.

The Copernicus program is coordinated and managed by the European Commission. The development of the observation infrastructure is performed under the aegis of the European Space Agency for the space component and of the European Environment Agency and the Member States for the in situ component.

Regarding Marine Monitoring Services, Copernicus provides regular and systematic reference information on the state of the physical oceans and regional seas. The observations and forecasts produced by the service support all marine applications. For instance, the provision of data on currents, winds and sea ice help to improve ship routing services, off-shore operations or search and rescue operations, thereby contributing to marine safety.

The service also contributes to the protection and the sustainable management of living marine resources in particular for aquaculture, fishery research or regional fishery organizations.

Physical and marine biogeochemical components are useful for water quality monitoring and pollution control. Sea level rise helps to assess coastal erosion. Sea surface temperature is one of the primary physical impacts of climate change and has direct consequences on marine ecosystems. As a result of this, the service supports a wide range of coastal and marine environment applications.

The products delivered by the Copernicus marine environment monitoring service today are provided free of charge to registered users through an Interactive Catalogue available at: **www.myocean.eu**

Link to the Copernicus Marine Monitoring Service: **http://copernicus.eu/pages-principales/services/marine-monitoring**

The Copernicus Emergency Management Service provides all actors involved in the management of natural disasters, man-made emergency situations, and humanitarian crises with timely and accurate geo-spatial information derived from satellite remote sensing and completed by available in situ or open data sources.

The mapping component of the service (GIO EMS - Mapping) has a worldwide coverage and provides the above-mentioned actors (mainly Civil Protection Authorities and Humanitarian Aid Agencies) with maps based on satellite imagery. The service started operations on 1 April 2012.

The products generated by the service can be used as supplied (e.g. as digital or printed map outputs). They may also be combined with other data sources (e.g. as digital feature sets in a geographic information system) to support geospatial analysis and decision making processes of emergency managers.

GIO EMS - Mapping can support all phases of the emergency management cycle: preparedness, prevention, disaster risk reduction, emergency response and recovery.

The service is provided free of charge in rush mode, for emergency management activities which require immediate response and non-rush mode, to support emergency management activities not related to immediate response (read more). It can be activated only by authorized users.

More information on the service is available on the GIO EMS – Mapping Portal:
http://emergency.copernicus.eu/mapping

Link to the Copernicus Emergency Management Service: **http://copernicus.eu/pages-principales/services/emergency-management**

Step 6: Sponsor the research and development of new technologies for identification, remote sensing, monitoring, assessment and elimination of sea-disposed chemical weapons.

Epilogue

"This is a problem which will not go away. It will have to be tackled by the nations it most concerns – the original manufacturers of the chemical weapons (Germany, Russia, United Kingdom, United States), the principal dumpers (Russia, United Kingdom, United States), and those countries most at risk."[139]

In his assessment of the dilemma, Mr. V. Sherbakov, Vice Mayor of St. Petersburg, Russia, stated: "Taking into consideration the vital importance of the problem, I'd like to emphasize that the wrong assessment of the situation or the delay of conclusions are intolerable because of the possible dramatic consequences."[140]

Author Fredrik Laurin warned: "No one knows what to do with the chemicals, or what the chemicals will do if left alone. The one certainty is that their containers are deteriorating. Ship hulls constructed early this century were 20-25 millimeters thick. Skagerrak seawater, with its high salt content, can break through such metal in 40 or 50 years. Leaking shells have already been retrieved."[141]

Yevgeniy Solomenko advised in the Mosscow Izvestiya: "The `Baltic Chernobyl' could strike the bell of disaster; we cannot wait. All operations in the Baltic must be completed. Otherwise the chemical shells might begin to become unsealed on a large scale."[142]

Bottom Line: "Once again the old saying holds true: no matter how you may try to conceal a thing, the truth will come out. Sooner or later all secrets are revealed. Now it is too late to lash out at the people who were responsible for this planned

environmental disaster in the Baltic Sea. No amount of angry words can help in this case. Now we need to do something, because there is little time left for decisive action."[143]

APPENDIX A

ABBREVIATIONS

AC	Hydrocyanic Acid / Hydrogen Cyanide (blood agent)
ACC	Allied Control Council
ACDA	Arms Control and Disarmament Agency, U.S. Department of State
ACofS	Assistant Chief of Staff
ACW	Abandoned Chemical Weapons
AEAF	Allied Expeditionary Air Forces
AF	Air Force
APG	Aberdeen Proving Ground, Maryland
ASP	Ammunition Supply Point
BAOR	British Army of the Rhine
BW	Biological Weapon, Biological Warfare
BZ	Incapacitating Agent: 3-Quinuclidinyl Benzilate (hallucinogenic)
ca.	Circa
CA	Bromobenzylcyanide
CADC	Continental Ammunition Dumping Committee (U.K.)
CADC(G)	Continental Ammunition Dumping Committee, Germany sub-group
CADC(UK)	Continental Ammunition Dumping Committee, U.K. sub-group
CAM	Chemical Agent Monitor
CAMDS	Chemical Agent Munitions Disposal System (U.S.)

CBDE	Chemical and Biological Defence Establishment (U.K.)
CCG(BE)	British Element of the Control Commission for Germany
CCmlO	Chief Chemical Officer
CCMS	Committee on the Challenges of Modern Society, NATO
CCWS	Chief, Chemical Warfare Servive
CDE	Chemical Defence Establishment (U.K.)
CEQ	Council on Environmental Quality (U.S.)
CG	Commanding General
CG	Phosgene (lung poison/choking agent)
CIA	Central Intelligence Agency (U.S.)
CinC	Commander in Chief
CK	Cyanogen Chloride (blood agent)
CL	Chlorine Gas
CM	Chemical Munitions
CM-IN	Classified message, incoming
Cml	Chemical
CmlO	Chemical Officer
CM-OUT	Classified message, outgoing
CN	Chloroacetophenone
CNB	Solution of chloroacetophenone in benzene and carbon tetrachloride
CNS	Chloracetophenone with benzene and carbon tetrachloride (tearing agent)
CO	Commanding Officer
CofS	Chief of Staff
CONUS	Continental United States
CS	o-chlorobenzylidene-malononitrile
CW	Chemical Weapon, Chemical Warfare
CWC	Chemical Weapons Convention

CWS	Chemical Warfare Service
CWSO	Chemical Warfare Service Officer
CX	Phosgene Oxime (blister agent)
DA	Department of the Army
DA	Clark I (Diphenylchloroarsine) (an arsenical)
DANC	A decontaminant
DC	Clark II (Diphenylcyanoarsine) (an arsenical)
DF	Methyl Phosphonic Diflouride (di-fluoro/liquid precursor used with isopropyl alcohol to form GB)
DJL	Davey Jones Locker, U.S. Operation
DM	Adamsite (diphenylamine chloroarsine) (an arsenical)
DoD	Department of Defense (U.S.)
DoE	Department of the Environment (U.K.)
DP	Diphosgene (lung poison/choking agent)
DPR	Disarmament Progress Report
EC	European Community
ED	Ethyldichloroarsine (an arsenical)
EEC	European Economic Community
EEIS	Enemy Equipment Intelligence Service of the Chemical Warfare Service
EIA	Environmental Impact Assessment
EIS	Environmental Impact Statement
EPA	Environmental Protection Agency (U.S.), Environmental Protection Act (U.K.)
ESA	Environmentally Sensitive Area
ETO	European Theater of Operations
ETOUSA	European Theater of Operations, U. S. Army
EU	European Union
FECOMZ	Forward Echelon, Communications Zone

FM	Field Manual
FS	Fuming Sulfuric Acid
G Agents[144]	Nerve gases tabun (GA), sarin (GB), and soman (GD)
G-1	Personnel section of divisional or higher staff, U.S.
G-2	Intelligence section of divisional or higher staff, U.S.
G-3	Operations section of divisional or higher staff, U.S.
G-4	Supply section of divisional or higher staff, U.S.
GA	Tabun (ethyl NN-dimethylphosphoramidocyanidate) (nerve agent)
GAF	German Air Force
GAO	General Accounting Office/Government Accountability Office, U.S. Congress
GB	Sarin (isopropyl methylphosphonoflouridate) (nerve agent)
GD Soman	(1,2,2-trimethylpropyl methylphosphonoflouridate)
GESAMP	Group of Experts on the Scientific Aspects of Marine Pollution
H	Mustard / Levinstein Mustard (blistering agent)
HC	Hexachlorethane
H/CC3	Mustard/CC3 (Hungarian)
HD	Distilled Mustard (dichlorodiethyl sulfide) (blister agent)
HE	High Explosive
HELCOM	Helsinki Commission

HELCOM CHEMU	Ad-Hoc Working Group on Dumped Chemical Munition of the Baltic Marine Environment Protection Commission of the Helsinki Commission
HG	Mercury
HL	Mustard-Lewisite Mixture (an arsenical)
HN	Nitrogen Mustard
H/PD	Mustard Phenyldichlorarsine
HQ	Headquarters
HS	Sulfur Mustard (Dichloroethyl Sulfide)
HT	Thickened Mustard (blister agent)
L	Lewisite (blistering agent) (an arsenical)
LCT	Landing Craft, Tank
LD50	Median Lethal Dose
LE-100	U.S. Code Name for Tabun (GA)
LMAB	London Munitions Assignment Board
MD	Methyldichloroarsine (an arsenical)
Mil	Military
MoD	Ministry of Defence (U.K.)
NACC	North Atlantic Cooperation Council
NATO	North Atlantic Treaty Organization
NBC	Nuclear, Biological, and Chemical
NDRE	Norwegian Defence Research Establishment, Norway
NDRETOX	Division for Environmental Toxicology of the Norwegian Defence Research Establishment
NEPA	National Environmental Policy Act of 1969, U.S. Congress
OCW	Old Chemical Weapons
OMGUS	Official Military Government of the United States

OPCW	Organisation for the prohibition of Chemical Weapons
PD	Phenyldichloroarsine (an arsenical)
P.L.	Public Law (U.S.)
Ppm	Parts Per Million
PRO	Public Record Office (U.K.)
PS	Chlorpicrin (lung poison/choking agent)
QL	Liquid precursor used with sulfur to form VX
RAF	Royal Air Force
R&D	Research and Development
ROV	Remote-Operated Vehicle
Rpt	Report
S-3	Operations officer or section of regimental or lower staff (U.S.)
SA	Arsine / Arsenic Trihydride
SHAEF	Supreme Headquarters, Allied Expeditionary Forces
SIPRI	Stockholm International Peace Research Institute, Sweden
SMAD	Soviet Military Administration in Germany
SO3	Sulphur Trioxide
TCM	Toxic Chemical Munition
TM	Technical Manual
UNCED	United Nations Conference on Environment and Development
USCWC	U.S. Chemical Warfare Committee
VV	Thickened Mustard
VX	Standard persistent nerve agent
Z/I	Zone of the Interior (Continental USA)

APPENDIX B
CHEMICAL AGENT CODES

AC	Hydrocyanic Acid / Hydrogen Cyanide (blood agent)
BZ	incapacitating agent
CG	Phosgene (lung poison/choking agent)
CA	Bromobenzylcyanide
CK	Cyanogen Chloride (blood agent)
CL	Chlorine Gas
CN	Chloroacetophenone
CNB	solution of chloroacetophenone in benzene and carbon tetrachloride
CNS	Chloracetophenone with benzene and carbon tetrachloride (tearing agent)
CS	o-chlorobenzylidene-malononitrile
DA	Diphenylchloroarsine (Clark I) (an arsenical)
DC	Diphenylcyanoarsine (Clark II) (an arsenical)
DANC	(a decontaminant)
DF	di-fluoro/liquid precursor used with isopropyl alcohol to form GB
DM	Adamsite (diphenylamine chloroarsine) (an arsenical)
DP	Diphosgene (lung poison/choking agent)
ED	Ethyldichloroarsine (an arsenical)
FS	Fuming Sulfuric Acid
GA	Tabun (ethyl NN-dimethylphosphoramidocyanidate) (nerve agent)
GB	Sarin (isopropyl methylphosphonoflouridate) (nerve agent)

GD	Soman (1,2,2-trimethylpropyl methylphosphonoflouridate)
H	Mustard / Levinstein Mustard (blistering agent)
HC	Hexachlorethane
H/CC3	Mustard/CC3 (Hungarian)
HD	Distilled Mustard (dichlorodiethyl sulfide)
HG	Mercury
HL	Mustard-Lewisite Mixture (an arsenical)
HN	Nitrogen Mustard
H/PD	Mustard Phenyldichlorarsine
HS	Sulfur Mustard (Dichloroethyl Sulfide)
HT	Mustard Thickened
L	Lewisite (blistering agent) (an arsenical)
MD	Methyldichloroarsine (an arsenical)
NB	Oleum
PD	Phenyldichloroarsine (an arsenical)
PS	Chlorpicrin (lung poison/choking agent)
QL	liquid precursor used with sulfur to form VX
SA	Arsine / Arsenic Trihydride
SO3	Sulphur Trioxide
VV	Thickened Mustard
VX	Standard Persistent Nerve Agent

APPENDIX C

DETAILED CHRONOLOGY OF EVENTS IN THE AMERICAN AND BRITISH ZONES AND US, UK, RUSSIAN, AND GERMAN CW SEA DISPOSAL OPERATIONS

1 Sep 39   World War II started with the attack of Poland by Germany. The war lasted five and a half years, devastated much of Europe, and killed 55 million people.

12 Jul 41   Moscow: Agreement between the United Kingdom and the Union of Soviet Socialist Republics:

His Majesty's Government in the United Kingdom and the Government of the Union of Soviet Socialistic Republics have concluded the present Agreement and declare as follows:

(1) The two Governments mutually undertake to render each other assistance and support of all kinds in the present war against Hitlerite Germany.

(2) They further undertake that during this war they will neither negotiate nor conclude an armistice or treaty of peace except by mutual agreement.

The present Agreement has been concluded in duplicate in the English and Russian languages. Both texts have equal force.

By authority of His Majesty's Government in the United Kingdom:
R. STAFFORD CRIPPS
His Majesty's Ambassador Extraordinary and Plenipotentiary in the Union of Soviet Socialistic Republics.

By authority of the Government of the Union of Soviet Socialistic Republics:
V. MOLOTOV
The Deputy President of the Council of People's Commissars and People's Commissar for Foreign Affairs of the Union of Soviet Socialistic Republics.

6 Jun 44	D-Day Invasion
20 Jul 44	Germany: A coup attempt, carried out mainly by German officers, failed. Hitler survived a bomb planted in his HQ.
11 Sep 44	France: Except for the ports of Boulogne, Calais, and Dunkerque (Dunkirk), the entire French channel was under Allied control.
12 Sep 44	UK: The London Protocol

The London Protocol, an agreement reached among the UK, US, Soviet Union, and France, was to exercise total control over Germany. Germany was to be divided into three occupation zones and Berlin into three sectors. Overall control would be exercised by an Allied Control Council (ACC) composed of three Commanders-in-Chief. (At the Yalta Conference, 4-11 February 1945, France was co-opted as the fourth controlling power and allocated its own occupation zone.)

15 Sep 44   Germany: Units of the U.S. V and VII Corps forces reached the SW frontiers of Germany. Except for a strip of land on the German border, Belgium and Luxembourg have been completely liberated.

4 Dec 44   UK: Standing Committee on War Material (SCWM) meeting #1

The first meeting of the Standing Committee on War Material of the Control Commission for Germany (British Element) was held at 1500 hours, Monday, December 4, 1944 in the sixth floor conference room, Norfolk House. Brigadier E. I. G. Griffith-Williams, DSO, MC (Land Forces Division) and Brigadier W. E. Van Cutsen, J.H.R.S., were in the Chair.[145] The Standing Committee on War Material convened 28 times between December 4, 1944 and March 21, 1946.

9 Dec 44   UK: Standing Committee on War Material (SCWM) meeting #2

The second meeting of the Standing Committee on War Material of the Control Commission for Germany (British Element) was held at 1500 hours, Saturday, December 9, 1944 in the sixth floor conference room, Norfolk House. Brigadier E. L. G. Griffith-Williams, DSO, MC (Land Forces Division) and Brigadier W. E. Van Cutsen, J.H.R.S., were in the Chair.[146] "Brigadier Griffith-Williams referred to WMDD/P(44)8 which the Committee had before it, regarding the foundation of a special sub-committee to consider the dumping of surplus enemy ammunition at sea. The general feeling of the Committee was that, although a sub-committee would probably be required to consider this matter, it would be unwise to set up a single sub-committee on a particular subject without first examining the extent to which the Standing Committee should operate through sub-committees and also the full field which each sub-committee would need to cover so as to prevent duplication or omissions."

The Committee agreed:

a. That it would need to establish various sub-committees.

b. That it would reserve to itself the definition and listing of war material, and delegate to the sub-committees detailed aspects of disposal within the Standing Committee's own terms of reference.[147]

The Standing Committee on War Material agreed to establish six sub-committees, to include the Ammunition, Explosives, and Chemical Warfare Substances Sub-Committee.[148]

14 Dec 44   UK: Standing Committee on War Material (SCWM) meeting #3

The third meeting of the Standing Committee on War Material of the Control Commission for Germany (British Element) was held at 1030 hours, Thursday, December 14, 1944 in the sixth floor conference room, Norfolk House.

20 Dec 44   UK: Standing Committee on War Material (SCWM) meeting #4

The fourth meeting of the Standing Committee on War Material of the Control Commission for Germany (British Element) was held at 1030 hours, Wednesday, December 20, 1944 in the sixth floor conference room, Norfolk House.

3 Jan 45   UK: Standing Committee on War Material
            (SCWM) meeting #5

The fifth meeting of the Standing Committee on War
Material of the Control Commission for Germany (British
Element) was held at 1430 hours, Wednesday, January 3,
1945 in Room 311, Norfolk House. Brigadier E. L. G.
Griffith-Williams, DSO, MC (Land Forces Division) and
Brigadier W. E. Van Cutsen, J.H.R.S., were in the Chair.
"The Chairman suggested that it would now be opportune
to invite the U.S. Group Control Council to send an
observer to the Committee's meetings, and to establish
close liaison with SHAEF Enemy War Materials Board.
The Russians would similarly be invited to send an
observer as soon as their planners arrived in London."[149]

10 Jan 45   UK: Standing Committee on War Material
             (SCWM) meeting #6

The sixth meeting of the Standing Committee on War
Material of the Control Commission for Germany (British
Element) was held at 1500 hours, Wednesday, January
10, 1945 in Room 311, Norfolk House. Brigadier E. L. G.
Griffith-Williams, DSO, MC (Land Forces Division) was
in the Chair. Also present for the first time, and
representing the U.S. Army, was Colonel H. S. Struble.
"The Chairman said he took this opportunity to welcome
Colonel Struble and other representatives of the U.S.
Group Control Council, who had accepted the invitation
of the Committee to attend as observers. He stressed the
need for continuity, so far as possible, of policy and
practice, throughout the post-hostilities and post-
surrender periods, and uniformity in the different zones
of occupation. The presence of members of the U. S.

Group Control Council would enable the Committee to keep in touch with American proposals for the post-war period, and it was hoped that in due course they would also have the benefit of the presence of a Russian representative.[150]

24 Jan 45   UK: Standing Committee on War Material (SCWM) meeting #7

The seventh meeting of the Standing Committee on War Material of the Control Commission for Germany (British Element) was held at 1500 hours, Wednesday, January 24, 1945 in Room 311, Norfolk House. Brigadier E. L. G. Griffith-Williams, DSO, MC (Land Forces Division) was in the Chair. Five representatives from the U.S. Group Control Council were also in attendance:

> U.S. Group Control Council (Germany)
> Colonel H. S. Struble, U.S. Army
> Lieutenant H. G. Hastings, U.S. Navy
> U.S. Control Group (Austria)
> Colonel H. H. Cartwright, U.S. Army
> Lt.Col. R. P. Hutt, U.S. Army
> Lieutenant R. V. Copeland, U.S. Navy

"There was some discussion as to the applicability of the words 'captured' and 'surrendered'. 'Captured' war material would be all that taken before the end of organized hostilities, and 'surrendered' war material would be the balance of all war material not already captured. War material would be "captured" only during the SHAEF period, and it was, therefore, questionable whether the Control Commission should concern itself about captured material."[151]

4-11 Feb 45 Ukraine: Yalta Conference

The Allied conference near Yalta, in the Ukraine, was a meeting between Allied leaders Sir Winston S. Churchill (1874-1965), Franklin D. Roosevelt (1882-1946), and Joseph Stalin (1879-1953). The purpose of the Conference was to determine overall Allied strategy and set up guidelines for what would happen following the war. It was determined that all German militarism should be destroyed, and that all war criminals would be dealt with swiftly. In addition, the discussion of how Germany would be governed was touched on, and the idea of zones of occupation was introduced. France was co-opted as the fourth controlling power and allocated its own occupation zone. Also, it was agreed that the USSR would declare war on Japan within 90 days of the fall of Germany. Finally, it was declared that there would be a United Nations meeting in San Francisco following the war. The only Allied intention which remained valid was that of terminating Germany's existence as an independent state but keeping the country intact.

7 Feb 45   UK: Standing Committee on War Material
              (SCWM) meeting #8

The eighth meeting of the Standing Committee on War Material of the Control Commission for Germany (British Element) was held at 1500 hours, Wednesday, February 7, 1945 in Room 311, Norfolk House. Brigadier E. L. G. Griffith-Williams, DSO, MC (Land Forces Division) was in the Chair. Representing the U.S. Group Control Council were Colonel W. R. Carter, U.S. Army and Lieutenant Colonel (LTC) Adrian St.John, U.S. Army.

14 Feb 45   UK: Standing Committee on War Material
             (SCWM) Special Meeting

A special meeting of the Standing Committee on War
Material of the Control Commission for Germany (British
Element) was convened at 1030 hours, February 14, 1945
in Room 126, Norfolk House. "The Chairman, Brigadier
E. L. G. Griffith-Williams (Army Division), stated that
one of the objects of the meeting was to stress the fact
that both the main standing committee and the sub-
committees were purely planning advisory bodies set up
in order to produce agreed Commission views to the
D.C.s. They were not executive bodies and this fact could
not be too strongly emphasized."[152]

21 Feb 45   UK: Standing Committee on War Material
             (SCWM) meeting #9

The ninth meeting of the Standing Committee on War
Material of the Control Commission for Germany (British
Element) was held at 1500 hours, Wednesday, February
21, 1945 in Room 104, Norfolk House. Brigadier A. M.
Craig (Naval Division) was in the Chair. Representing
the U.S. Group Control Council were Colonel H. S.
Struble, U.S. Army and Lieutenant F. G. Hastings, U.S.
Navy.

7 Mar 45   UK: Standing Committee on War Material
             (SCWM) meeting #10

The tenth meeting of the Standing Committee on War
Material of the Control Commission for Germany (British
Element) was held at 1500 hours, Wednesday, March 7,
1945 in Room 507, Norfolk House. Brigadier E. L. G.
Griffith-Williams, DSO, MC (Land Forces Division) was

in the Chair. Representing the U.S. Group Control Council were Colonel H. S. Struble, U.S. Army and Lieutenant F. G. Hastings, U.S. Navy.

21 Mar 45   UK: Standing Committee on War Material (SCWM) meeting #11

The 11th meeting of the Standing Committee on War Material of the Control Commission for Germany (British Element) was held at 1500 hours, Wednesday, March 21, 1945 in Room 507, Norfolk House. Brigadier E. L. G. Griffith-Williams, DSO, MC (Land Forces Division) was in the Chair. Representing the U.S. Group Control Council was Lieutenant F. G. Hastings, U.S. Navy.

That the U.S. Group Control Council only dispatched a Navy Lieutenant as its representative is indicative of a desire to no longer be an active participant. As it turns out, this was, in fact, the last SCWM meeting attended by the U.S. Group Control Council. Of the 28 meetings held by the SCWM between December 4, 1944 and March 21, 1946, representatives of the U.S. Group Control Council only attended 6 (or 21.4%), meetings six through eleven. SHAEF, however, was represented for the first time by a Colonel L. G. Turnbull.

During the course of the meeting, the procedure for collating Allied requirements for German war material was discussed. "The instrument for collating British and U.S. requirements during the pre-defeat period was the London Munitions Assignment Board (LMAB). These requirements were then notified to SHAEF. During the post-defeat period, the LMAB would no longer be functioning, and there would virtually be a standstill on most war material requirements pending four-party agreement."

4 April 45   UK: Standing Committee on War Material (SCWM) meeting #12

The 12th meeting of the Standing Committee on War Material of the Control Commission for Germany (British Element) was held at 1500 hours, Wednesday, April 4, 1945 in Room 126, Norfolk House. Brigadier E. L. G. Griffith-Williams, DSO, MC (Land Forces Division) was in the Chair. No representative from the U.S. Group Control Council attended.

18 Apr 45   UK: Standing Committee on War Material (SCWM) meeting #13

The 13th meeting of the Standing Committee on War Material of the Control Commission for Germany (British Element) was held at 1500 hours, Wednesday, April 18, 1945 in Room 126, Norfolk House. Brigadier E. L. G. Griffith-Williams, DSO, MC (Land Forces Division) was in the Chair. The U.S. Group Control Council was again not represented.

19 Apr 45   American Zone: U.S. forces assumed control of Grafenwöhr Chemical Depot on April 19, 1945.

The third of five such depots located in the American Zone, Grafenwöhr remained operational for 695 days. Chemical officers from the Third U.S. Army conducted a preliminary survey of the Grafenwöhr storage area on April 24th and a preliminary inventory was conducted before January 1946. It was not until the 18th Chemical Maintenance Company arrived in February, 1946 that a detailed inventory was conducted. On March 15, 1947, the depot was closed and the 18th Chemical Maintenance Company was deactivated.

There were no ammunition bunkers, warehouses, or storage buildings of any kind in the Grafenwöhr Depot when the Americans arrived. All ammunition was piled in the open, along roadways, and in the woods. The dump consisted of approximately 2,000 acres of wooded, rolling land located about 3/4 miles NE of Grafenwöhr on the road between that town and Pechof. No buildings other than two small one-room hunting cabins were in the area. No storage of decontamination or protective equipment was found and no German civilian or military personnel who were familiar with the storage could be located.

The munitions in storage consisted of approximately 2,000,000 rounds of artillery shells, Nebelwerfer rockets, landmines, and approximately 75% of the storage is of German manufacture and design; the other 25% consisted of captured French, Hungarian, and a small lot of Russian and Czechoslovakian shells. Calibres varied from 77 mm to 210mm, the greatest proportion being 105 mm. Dates of manufacture extended from 1918 (French phosgene shells) to February 1945 (Nebelwerfer green ring rockets).

21 Apr 45   Russian Zone: The First White Russian Front was rapidly advancing north towards Berlin.

22 Apr 45   Germany's Heinrich Himmler made a surrender offer of the German Army to the western Allies, though not to the Russians.

24 Apr 45   American Zone: American chemical officers

of the Third U.S. Army made preliminary surveys of the Grafenwöhr storage area. No chemical company was assigned to the depot before the arrival of the 18th Chemical Maintenance Company in February, 1946. A preliminary inventory, however, was taken by the 2d Chemical Mortar Battalion before January 1946.

25 Apr 45   UK: Ammunition Dumping experiment in St. Catherine's Deep.

The following is a transcript of the "Report of Ammunition Dumping Trials by L.C.T (4)," dated April 26, 1945:

From: Flotilla Officer, 37th L.C.T. Flotilla
To: Maint. Captain, Landing Ships and Craft, H.M.S. Squid, Southampton

"On Wednesday, 25th April 1945, L.C.T. 937 with 194 tons of ammunition on board slipped from 38 berth at 0830 and proceeded to St. Catherine's Deep to carry out experiments in ammunition dumping by conveyor belts. Protective clothing was worn during the trials. L.C.T. 937 was escorted by H.M.T. Tamora and American Officers in tug S.T. 769 were in company as spectators.

When abeam of "F" Buoy at 1122, the senior Army Officer was informed that dumping of the ammunition would commence in approximately one hour and suggested that the Army personnel should have their dinner now, which was concurred in. As it was noticed that this meal consisted of one cheese sandwich, a small piece of cake, and one biscuit per man, it was suggested to the senior

Army Officer that in view of the heavy work ahead, this meal could be augmented, if desired, from emergency stores carried on board, and that tea, sugar, and milk could be issued as well. This was accepted and appreciated by all concerned.

Dumping of the ammunition commenced at 1220 in position 50°35'10" N, 1°9' W. The experiment appeared to be a complete success with the exception of certain boxes containing phosgene bombs. These were rather slow in sinking but when extra holes had been made in the boxes, this fault was found to be remedied. Dumping was completed at 1455 when in position 50°34'15" N, 1°12'30" W. L.C.T. 937 then returned to Southampton and was made fast on 36 berth by 1925."

27 Apr 45	Berlin: The Allied reply to Heinrich Himmler's April 22, 1945 offer of surrender was delivered by Count Bernadotte. The answer was "nothing short of unconditional surrender on all fronts".
27 Apr 45	Berlin: The Russians now control 75% of Berlin.
29 Apr 45	American Zone: U.S. forces assumed control of Schierling Chemical Depot on April 29, 1945. The fourth of five such depots located in the American Zone, Schierling remained operational for 711 days. On April 10, 1947, the depot was closed and the 140th Chemical General Service Company was deactivated.

30 Apr 45   Adolf Hitler committed suicide in the
             Chancellery Bunker, Berlin.

May 45   By May 1945 no clear agreement had been
             reached by the British, American, and
             Russian allies as to what was covered by the
             term "war material" in the context of the
             demilitarization of Germany. As a result, the
             British Element of the Control Commission
             for Germany [CCG(BE)], the agency
             responsible for administering the British
             Occupied Zone of Germany, noted that there
             was unlikely to be anything approaching a
             uniform treatment of German war material
             throughout all the Zones (British, American,
             French, and Russian).[153]

May 45   British Zone: BAOR Disarmament Progress
             Report #1

The British Army of the Rhine (BAOR) prepared its first
periodic Disarmament Progress Report to report on
demilitarization activities in the British Zone.

2 May 45   UK: Standing Committee on War Material
             (SCWM) meeting #14

The 14th meeting of the Standing Committee on War
Material of the Control Commission for Germany (British
Element) was held at 1500 hours, Wednesday, May 2,
1945 in Room 216, Norfolk House. Brigadier E. L. G.
Griffith-Williams, DSO, MC (Land Forces Division) was
in the Chair. No representative from the U.S. Group
Control Council was in attendance.

2 May 45   US: Declassified SECRET Transcript of
            Telephone Conversation between Major
            General Porter (Chief, Chemical Warfare
            Service) and Brigadier General Alden H.
            Waitt, Assistant Chief for Field Operations,
            Chemical Warfare Service.

3 May 45   American Zone: U.S. forces assumed control
            of St. Georgen Chemical Depot on May 3,
            1945. Despite being the last of five such
            depots to surrender to U.S. forces, St.
            Georgen remained operational longer than
            any other depot in the American Zone - 759
            days. On June 1, 1947, the depot closed and
            the 193rd Chemical Depot Company, which
            had been in charge of depot activities since
            January 1946, was deactivated.

German Disposal Operations
In the last days of the war, two ships with 69,000 tabun
shells and 5,000 tons of other chemical munitions were
disposed in the Baltic's Little Belt.

8 May 45   V-E Day (Victory in Europe Day)

The unconditional surrender of Germany was formally
ratified and confirmed in Berlin ending the war in
Europe.

11 May 45   UK: Standing Committee on War Material
            (SCWM) meeting #15

The 15th meeting of the Standing Committee on War
Material of the Control Commission for Germany (British
Element) was held at 1500 hours, Friday, May 11, 1945
in Room 126, Norfolk House. Brigadier E. L. G. Griffith-
Williams, DSO, MC (Land Forces Division) was in the
Chair. No representatives from the U.S. Group Control
Council were in attendance.

16 May 45   UK: SECRET Letter (now declassified) from
            the War Office regarding the dumping of
            U.S. munitions in Beaufort's Dyke and the
            dumping of conventional and toxic U.K.
            munitions in Beaufort's Dyke and the
            English Channel:

"The problem divides itself into two parts:

Firstly, the disposal of 958 tons of 150 lb. bombs
belonging to the U.S. ARMY from the vessels "Bantom"
and "Cape Borda" which have been waiting in U.K.
waters from 17 April. Arrangements have now been made
for these two vessels to be dealt with at 'LOCH NA
KEAL' where they will arrive on 14 May. The bombs will
be transferred to 2 L.C.T.s provided by the Admiralty and
manned under military arrangements. The craft will then
proceed to 'BEAUFORT DYKE' the dumping ground for
the ammunition.

Secondly, there is in this country approximately 50,000
tons of unserviceable ammunition which includes a
portion of toxic ammunition to be disposed of by dumping
at sea. The ammunition immediately available is 19,803

tons of which 14,607 tons are located in the North and 5,196 tons are located in the South. The ports suggested for the loading of L.C.T.s are CAIRN-RYAN and/or WORKINGTON for final dumping in the BEAUFORT DYKE and loading at Little Hampton and/or New Haven for final dumping in the English Channel." (See the 7 Jun 45 message from the Admiralty to the Commander-in-Chief, Portsmouth.)

24 May 45   US: Declassified SECRET Radio CM-OUT 87411, from the Assistant Chief for Material, Chemical Warfare Service to Headquarters Communications Zone, European Theater of Operations (ETO), U.S. Army Paris.

Jun 45   British Zone: BAOR Disarmament Progress Report #2

The second BAOR Disarmament Progress Report noted that among the priorities at that time was removal into guarded dumps of all munitions, including chemical munitions. As part of this work, a detailed recon program was already in hand. The report also noted that at that time, destruction of enemy war material was forbidden, except for security reasons, pending a decision from the Potsdam Conference in July/August 1945.

6 Jun 45   US: Canadian Chemical Warfare Laboratory Report No. 44

7 Jun 45   UK: Message 072159B/June sent at 2159 hours June 7, 1945 from the Admiralty to the Commander-in-Chief, Portsmouth regarding the disposal of conventional and toxic ammunition in the English Channel.

"1. War Office desire to dispose of 5,196 tons of unserviceable ammunition including a portion of toxic ammunition by dumping at sea in English Channel.

2. It is proposed that two L.C.T. (4) should be lent to Army and manned under military arrangements, and that craft should load at Littlehampton or New Haven for final dumping in English Channel.

3. Request you will specify area after consultation with local fishery office."

See the 10 Jun 45 response to the Admiralty from the Commander-in-Chief, Portsmouth.

8 Jun 45   UK: Standing Committee on War Material (SCWM) meeting #16

The 16th meeting of the Standing Committee on War Material of the Control Commission for Germany (British Element) was held at 1500 hours, Friday, June 8, 1945 in Room 311, Norfolk House. Brigadier E. L. G. Griffith-Williams, DSO, MC (Land Forces Division) was in the Chair. No representative from the U.S. Group Control Council attended. This was the last meeting of the SCWM to be held in England. Allowing two months for the transfer, the Committee reconvened in Bad Oeynhausen, Germany on August 7, 1945. The final 12 meetings of the Committee were held there.

10 Jun 45   UK: Message 101747B/June sent at 1747 hours June 10, 1945 from the Commander-in-Chief, Portsmouth to the Admiralty regarding the disposal of conventional and toxic ammunition in the English Channel:

"1. The proposal in A.M. 072159 (your message of 7 June) is not concurred in.

2. St. Catherine's Deep has, of necessity, been used for dumping small quantities of unserviceable ammunition as a wartime measure. In consideration of fishing interests, however, this practice should now be strictly limited to emergency dumping of small quantities. The use of this area for dumping large quantities is unacceptable.

3. It is considered that the only area in the English Channel where large amounts of ammunition can be dumped without prejudice to fishing interests is the Hurd Deep.

4. As ammunition of toxic type is liable to cause damage in fishing grounds over a wide area, it is considered that large quantities of this should not be dumped anywhere in the English Channel. The trials of dumping a small quantity of toxic ammunition in St. Catherine's Deep and the disposal of defective American ammunition elsewhere in April 1945, were only agreed to due to war conditions."

13 Jun 45   US: CONFIDENTIAL Voyage Report (now declassified) from Lieutenant (Junior Grade) Walter P. Halstead, U.S. Navy Reserve, the Commanding Officer, Naval Armed Guard aboard the S.S. Cape Borda to the Chief of Naval Operations. This report validates the information contained in the 16 May SECRET letter from The War Office cited above. The following is the "General Resume of Voyage":

"The ship arrived in Belfast Lough, N. Ireland at 2150 hours, 19 April 1945. While enroute it was discovered by authorities ashore that the 150 lb bombs carried by the ship were defective and liable to explosion upon any slight disturbance. In addition to the 620 tons of defective bombs, the ship also carried approximately 3,000 tons of detonators and other explosives. While waiting in Belfast for instructions concerning the proper disposal of this cargo, the hostilities in Europe were officially ended.

In order to minimize the destruction that might be caused to surrounding areas in case of an explosion during unloading, the ship was moved 13 May 1945 (departing at 0700 hours) to a remote Loch (Loch NaKeal) on the Isle of Mull, Scotland (arriving at 1905 hours). The unloading (which did not actually start until 18 May) was done at anchorage by a mixed detachment of volunteers from the Royal Engineers and the U.S. Army.

The ship arrived in Belfast Lough on 19 April 1945 and did not start unloading cargo until 18 May 1945. This long period of inactivity would indicate an unusual lack of decision and co-ordination on the part of persons responsible for the disposal of this dangerous cargo.

At 0530 hours, 27 May, upon completion of unloading the defective bombs, the ship proceeded to Clyde, Scotland (arriving at 1930 hours the same day) where it filled empty cargo space with 50-calibre machine gun ammunition. The ship was then routed to James Watt Docks, Cartsdyck, Scotland to load additional cargo (departing Clyde at 1455 hours 30 May).

The Cape Borda left James Watt Docks, Scotland at 1940 hours 3 June 1945 bound for the U.S., arriving in Philadelphia, Pennsylvania on 13 June 1945."[154]

In summary, a total of 958 tons of 150 lb. bombs belonging to the U.S. Army were transferred to the British Army at Loch NaKeal, Isle of Mull, Scotland (620 tons from the Cape Borda and 338 tons from the Bantom). It is concluded that the bombs were not "defective" at all; rather, when it became obvious the war was coming to an end, U.S. authorities decided the bombs were no longer needed and sought British assistance in their immediate disposal. The bombs were transferred to two L.C.T.s provided by the Admiralty and dumped in Beaufort's Dyke.

25 Jun 45   UK: First sub-committee meeting of the C.M.S.C. regarding unserviceable ammunition to be dumped at sea:

"At the Sub-Committee meeting of the C.M.S.C. held on 25th June 1945, the War Office asked if ships could be provided to act as safety vessels and mark ships in the dumping areas in Beaufort's Dyke and Hurd Deep.

The dumping in Beaufort's Dyke has been discussed with Commander-in-Chief W.A. who does not consider that such a ship is necessary, as the dumping ground is so close to the shore.

In view of the urgency of disposing of the 30,000 tons of the ammunition in the Channel Islands it is, however, considered that L.C.T. will have to be used for dumping as no other craft are available."[155]

30 Jun 45   UK: CCG(BE) Progress Report #12. The 12th progress report of the Control Commission for Germany (British Element) covering the period 1-30 June, 1945. (TOP SECRET document now declassified)

"The main concern and work of the British Element during the month under review was directed towards the study and detailed preparation of the move of the Control Commission to Germany."[156]

Jul 45   British Zone: BAOR Disarmament Progress Report #3

With specific regard to chemical munitions, the BAOR's July 1945 report noted that "it is believed that all major depots have now been discovered, the contents of the small dumps have been removed to the larger dumps (all of which are guarded), and there is a total of approximately 111,000 tons of enemy CW munitions in the British Zone."[157] Regarding disposal options, the BAOR report also noted that considerable thought had been given to the problem of finding the most economical and acceptable method of disposal. Of the three methods considered (dumping at sea in deep water, interring in deep and flooded mines, or industrial break-down), sea disposal was considered necessary for the bulk of the munitions in question.

2 July   First of 24 UK Atlantic Disposal Operations (1 ship: Empire Fal)

4 Jul 45   US: Declassified Memorandum from the Assistant Chief, CWS to the Chief, Theaters Branch: Packing of Captured German Toxics

7 Jul 45   UK: Message 071928B/June sent at 1928 hours July 7, 1945 from the Admiralty to the Commander-in-Chief, Plymouth (not Portsmouth) regarding dumping in Hurd Deep.

"From: Admiralty
To: C.inC. Plymouth and N.O.I.C., Channel Islands.
1. War Office has 30,000 tons of German ammunition in Channel Islands, which it is intended to dump in Hurd Deep. Three L.C.T. (4) manned by Army crews will be used for this purpose about 10th July.
2. The disposal of unserviceable ammunition from UK by dumping in Hurd Deep from L.C.T. (4) working from Littlehampton is also under consideration."

9 Jul 45   UK: Message 091434/July sent at 1434 hours July 9, 1945 from the Commander-in-Chief, Plymouth to the N.O.I.C. Channel Islands regarding dumping in Hurd Deep.

"From: C.inC. Plymouth
To: N.O.I.C. Channel Islands
Info: Admiralty, War Office, and C.inC. Portsmouth
IMPORTANT
1. (Reference your message of 7 July) A.M. 071928B. Request you will co-ordinate this operation (SEARAIL 7's message 021730B also refers).
2. The ammunition is to be dumped in the following positions in rotation:

(A) 49°38'00" N, 03°05'00" W
(B) 49°39'00" N, 03°01'00" W
(C) 49°40'30" N, 02°51'30" W"

9 Jul 45   US: Declassified SECRET Message (Radio CM-OUT 29437) from the Chemical Warfare Service to HQ, Communications Zone, European Theater of Operations (ETO), Paris.

11 Jul 45   UK: Second Sub-Committee meeting of the C.M.S.C. regarding unserviceable ammunition to be dumped at sea.[158]

15 Jul 45   British Zone: Continental Ammunition Dumping Committee (CADC) Meeting #1.

The Continental Ammunition Dumping Committee (CADC) held its first meeting on July 15, 1945 at HQ, 21st Army Group. Colonel C.H. Wooll, Q(M), 21st Army Group, was in the Chair. The CADC was established by the Service authorities in London to centrally control the task of ammunition disposal and in particular to organize the dumping at sea of captured enemy ammunition not required for research or to meet the requirements of the London Munitions Assignment Board (LMAB[159]). (The LMAB was established to allocate specific items of German munitions, usually for research purposes, among the various members of the United Nations).

Although there was close liaison with the American forces, and that information was passed to the Norwegian, Danish, and Dutch authorities, the CADC was not an inter-Allied body. It was a British organization established solely to deal with ammunition in the British Zone; stocks under British control in Belgium, the Netherlands, and Denmark; and certain stocks of surplus British ammunition which it was desired to dump.

Note that the CADC was sub-divided into the CADC (Germany), which assumed responsibility for operations on the continent and the CADC (UK) which handled the disposal of surplus British CW. Both the CADC (Germany) and the CADC (UK) reported to the London-based Continental Movements and Shipping Committee. According to the MoD, there is no evidence of any discussions of the CADC's program with the Russian authorities.[160]

It was agreed that the Army, through the 21st Army Group, would prepare the ammunition dumping program, including the program for CW munitions disposal. Note that this program only concerned dumping where the use of scuttled ships was concerned; the CADC was not responsible for local arrangements for the use of small craft such as lighters for the dumping of loose conventional ammunition. It was also decided at this meeting that "suitable" sites for the disposal of mustard gas (H) munitions included an area off Stavanger [161] Norway at a depth of 100 fathoms (only 183 meters) and in the Kattegat [162]. No decision, however, could be made without further clearance from London.[163]

16 Jul 45   UK: Message 161103/July sent at 1103 hours July 16, 1945 from the N.O.I.C. Channel Islands to the Commander-in-Chief, Plymouth regarding captured German ammunition to be dumped in U.K. waters.

"From: N.O.I.C. Channel Islands
To: C.inC. Plymouth
Info: C.inC. Portsmouth, Force Com. 135 Admiralty, War Office

(Reference your message of 7 July) A.M. 071928/July. Now estimated that in Channel Islands there are between 60,000 and 70,000 tons of German ammunition to be dumped."

17 Jul 45   Germany: The Potsdam Conference

President Truman, Prime Minister Winston Churchill, and Premier Josef Stalin began a conference in Potsdam, near Berlin, with Truman presiding. At this meeting the four powers agreed on the matters of denazification, demilitarization, economic decentralization, and the reeducation of the Germans along democratic lines. The conference lasted until 2 Aug 45.

18 Jul 45   US: Declassified SECRET Message (Radio CM-OUT 34527) from the Chemical Warfare Service to HQ, Communications Zone, US Forces European Theater, Paris, France.

25 Jul 45   Potsdam: After the 25 July 1945 session, the conference recessed to allow Prime Minister Churchill to return to England to obtain election results.

28 Jul 45   UK: Prime Minister Winston Churchill resigned with the Labour Party's assumption of power. Clement R. Attlee assumed responsibilities as Prime Minister.

31 Jul 45   Potsdam: The Potsdam Conference resumed with Prime Minister Clement R. Attlee representing the U.K.

31 Jul 45   British Zone: CCG(BE) Progress Report #13

The 13th progress report of the Control Commission for Germany (British Element) covering the period 1-31 July, 1945. The following information concerning CW disposal was excerpted from Appendix A, Naval Division Report:

"The dumping of CW ammunition and stores has been the subject of separate consideration. The Ministry of Agriculture and Fisheries has agreed to dumping in waters of not less than 300 fathoms and an area has been established in the Skagerrak. It is proposed to make use of certain hulks which are beyond repair so as to economise in shipping; the hulks to be scuttled with their cargo of 'C.W.' material."

The following table lists captured German warships, destroyers, and torpedo boats, which could potentially be used for CW dumping operations.

NAVAL DIVISION PROGRESS REPORT
"Summary of Progress from the Date of German
Capitulation Up Till 31st July, 1945"

GERMAN WARSHIPS

SHIP TYPE	NAME	LOCATION	DISPOSITION
6" Cruisers	Nurnberg	Wilhelmshaven	
	Leipzig	Wilhelmshaven	
	Köln	Wilhelmshaven	Sunk
	Emden	Kiel Fiord	Stranded & Gutted
	Berlin	Kiel	Ineffective
Pocket Battleships	Lutzow	Swinemunde	Damaged
	Admiral Scheer	Kiel	Capsized
8" Cruisers	Hipper	Kiel	Sabotaged in dock
	Prinz Eugen	Wilhelmshaven	
Old Battleships	Schlesien	Kaiser Canal	
Old Light Cruiser	Amazone	Kiel	Ineffective

DESTROYERS AND TORPEDO BOATS

LOCATION	DESTROYERS	TORPEDO BOATS
Wilhelmshaven	T-23, T-33, Z-5, Z-6, Z-10, Z-14, Z-15, Z-20, Z-25, Z-33, Z-34	F-2, F-4, F-7, F-8, F-10, T-14, T-17, T-19, T-107, T-108, T-151, T-153, T-158, T-190, T-196, TF-2
Brunsbuttel	Z-31	T-2, T-4, T-11, T-12
Frederikshaven		T-16
TOTAL:	12	22

"Two "Narvik" class destroyers and two "Elbing" class destroyers were sailed to the U.K. on 7th July, one of each being required for examination by the Admiralty and U.S. Navy, respectively."[164]

Aug 45 UK: The British marine attaché announced to the Danish marine ministry the intention of dropping large amounts of chemical weapons in international waters 15-20 nautical miles south of the Norwegian coast between 8°45' and 10°40' East [165]. The depth in that area is at least 450 meters (hereafter referred to as the "larger area").

1 Aug 45   British Zone: Continental Ammunition
             Dumping Committee (CADC) meeting #2

The second meeting of the Continental Ammunition
Dumping Committee (Germany) or CADC(G) took place
at HQ, 21st Army Group, with membership now
expanded to include a representative from the U.S.
Army's Port Command at Bremen, Germany.[166] The
details of the proposed dumping sites off Norway and
Denmark were to be passed to the Committee by the
Naval Commander-in-Chief Germany so they could be
relayed to the Admiralty and the Ministry of Agriculture
and Fisheries. An area in the Skagerrak [167] was
considered especially suitable as the depth of water there
was more than 300 fathoms (549 meters).

2 Aug 45   Potsdam: The Potsdam Conference ended
             (17 July-2 August 1945)

The Berlin Protocol of August 2, 1945 set out guidelines
for the "complete disarmament and demilitarization of
Germany" and further stated that "all arms,
ammunition, and implements of war and all specialised
facilities for their production shall be held at the
disposal of the Allies or destroyed." [168] This agreement
formed the basis for the dumping of chemical weapons
which took place after the war. [169] The detailed
discussions of this subject were delegated to the Inter-
Allied "Allied Control Commission" and its various
subcommittees, the overall controlling body being the
Allied Control Council headquartered in Berlin. In the
secret protocols of the Potsdam Conference, another
inter-Allied body was created to facilitate the
demilitarization process – the Tripartite Naval

Commission (US, UK, and USSR). Its task was to organize the disposal of the vessels of the German fleet. The Commission's decisions on these disposals required in many cases the scuttling of German naval vessels in deep water. Some could be used for the CW dumping program, but others which were not were in some instances scuttled within the areas used for CW-laden hulks, but not themselves carrying any CW munitions.[170]

2 Aug 45	US: Declassified Air Mail Letter from Brigadier General Alden H. Waitt, Assistant Chief for Field Operations, Chemical Warfare Service to the Chief Chemical Warfare Officer, HQ, Communications Zone, US Forces European Theater (Rear), Subject: Shipment of German Chemical Warfare Items not Available in USA.
7 Aug 45	British Zone: Standing Committee on War Material (SCWM) meeting #17

The 17th meeting of the Standing Committee on War Material of the Control Commission for Germany (British Element) was held at 1500 hours, Tuesday, August 7, 1945 in the Chief of Staff's Conference Room, Koenigshof Hotel, Bad Oeynhausen, Germany. This was the Committee's first meeting in Germany. For the first time, Brigadier K. R. Brazier-Creagh (Army Division) was in the Chair, vice Brigadier E. L. G. Griffith-Williams. And, despite the fact the Committee had moved to Germany, no U.S. representatives attended.

14 Aug 45	US: Letter from Colonel Delancey R. King, Chemical Warfare Service to the Commanding General, Army Service Forces.

16 Aug 45   British Zone: Standing Committee on War
Material (SCWM) meeting #18

The 18th meeting of the Standing Committee on War
Material of the Control Commission for Germany (British
Element) was held at 1500 hours, Thursday, August 16,
1945 in the Chief of Staff's Conference Room, Koenigshof
Hotel, Bad Oeynhausen. In lieu of Brigadier K. R.
Brazier-Creagh, Colonel M. St. J. Oswald (Army Division)
was in the Chair. Again, no U.S. representatives were in
attendance.

Priorities for dumping ammunition were discussed at this
meeting: "The Chairman suggested that the Committee
should obtain information regarding the programme for
dumping being worked out by the Continental
Ammunition Dumping Committee (CADC). Lt.Col. S. H.
M. Battye (Army Division) stated that so far as Army
Division was concerned, the only ammunition requiring
immediate priority was chemical warfare (CW)
ammunition. Air Commodore F. N. Trinder stated that
Air Division had large quantities of CW bombs which
they were anxious to dispose of as soon as possible.
Captain R. H. Bevan, R.N., stated that the Naval
Division was not immediately concerned to find dumping
facilities for any particular type of ammunition, but that
with regard to any programme for dumping, it seemed to
him that transportation and movement would be
overriding considerations.

During the discussion that followed, it was stated that all
Service Divisions were represented on the Continental
Ammunition Dumping Committee of which Colonel C. H.
Wooll was the Chairman." [171]

16 Aug 45   British Zone: Continental Ammunition
             Dumping Committee (CADC) Meeting #3

The Continental Ammunition Dumping Committee held
its third meeting at HQ, 21st Army Group on August 16,
1945. Colonel C.H. Wooll, Q(M) Shipping, 21st Army
Group, was in the Chair. The minutes from this meeting
are most significant because Appendix "A" to the minutes
contain the first details of the British Army's "C.W.
Ammunition Dumping Programme." (Depot numbers,
locations, tonnage, loading ports, and shipping were
allocated.)

17 Aug 45   In the absence of policy guidance or direction
             from the Berlin-based Allied Control
             Council, the Combined Chiefs of Staff (i.e.,
             the American and British Chiefs of Staff)
             advised the respective commanders of the
             British and American Zones in Germany
             that in the absence of any agreement by the
             Allied Control Council, both Zonal
             Commanders could investigate their own
             measures to destroy German war material
             in their respective areas.[172]

20 Aug 45   Norwegian authorities (the Norwegian
             Directorate of Fishing, in particular)
             objected to this large area, but agreed to a
             smaller area 21-24 nautical miles south of
             the coastline between 9°27' and 9°40' East.
             The depth in that area is greater than 650
             meters.

The official message to the CADC on August 20, 1945 read:

From: Searail 7
To: Chairman, Continental Ammunition Dumping Committee

"Following for quartermaster general Norwegian Army begins ref No J nr 3758/45 JR/SK. The Director of Fisheries maintains that the ammunition ought to be taken out and dumped outside the continental platform either in the Atlantic ocean or north of the North-sea alternatively in Skagerrak within an area enclosed by lines joining the following positions:

> A   58 DEGS 14 MINS N 09 DEGS 27 MINS E
> B   58 DEGS 16 MINS N 09 DEGS 27 MINS E
> C   58 DEGS 19 MINS N 09 DEGS 40 MINS E
> D   58 DEGS 17 MINS N 09 DEGS 40 MINS E

Should this area be chosen, the area must be marked with buoys beforehand. The Ministry can not agree to the dumping in any other place in Norwegian waters than those mentioned by the Director of Fisheries." [173]

At the same time, "Store Nordiske Telegrafselskap" objected to the proposed dumping `because the main telephone cable between Norway and Hirtshals, Denmark went through the so-called larger area. Norwegian authorities also stated that all ammunition dumped should also be stored in ships.

23 Aug 45   British Zone: CONFIDENTIAL Letter (now declassified) from the Commander-in-Chief, 21st Army Group to The Under Secretary of State and The War Office, Subject: Disposal

of Chemical Ammunition. This letter is significant in that it provides an inventory of captured German chemical weapons at 14 depots as of 14 Aug 45 and reveals plans to augment the British CW stockpile with captured German munitions charged with Tabun (GA).

"Details of CW ammunition taken under control up to 14 Aug 45 are given at Appendix A (see Appendix 4.3). In addition to this there are 4,000 tons of bulk mustard/aerosol in Hanover Province (30 Corps District). Details are also given in 21st Army Group Disarmament Progress Reports.

Dumping this ammunition in the sea is to start shortly. AMSSO Top Secret Signal 5151 of 171730B from Chiefs of Staff authorises the destruction of enemy war material with certain exceptions. No Grünring 3 natures are to be dumped for the moment as this charging is in short supply. Firm demands for any large amounts of enemy CW ammunition required in the UK are awaited.

The disposal of 4,000 tons of bulk chemical agent will not be attempted until advice is received on a reliable method of destruction." [174]

29 Aug 45   British Zone: Continental Ammunition Dumping Committee (CADC) Meeting #4

The Continental Ammunition Dumping Committee held its fourth meeting at HQ, British Army of the Rhine (BAOR) on August 29, 1945. Colonel C.H. Wooll, Q(M) Shipping, was in the Chair. An update to the "CW Ammunition Loading Programme" first presented in the minutes of the 16 August meeting is included as Appendix C.[175] As of August 1945, 110,643 tons of chemical warfare ammunition were scheduled for sea disposal. According to Appendix E of the minutes, The Royal Navy was responsible for all arrangements for the scuttling of ships.

31 Aug 45   British Zone: CCG(BE) Progress Report #14

The 14th progress report of the Control Commission for Germany (British Element) for August 1945. The following information concerning CW disposal was excerpted from Appendix A, Naval Division Progress Report for Period 31st July - 31st August 1945:

"In last month's report it was mentioned that CW material would be disposed of by scuttling old ships laden with it in deep water in the Skagerrak. 34 ships have been allocated and loading will start during the 1st week of September at Kiel (Nordhafen), Emden, Schlutup, and Flensburg. This programme allows for the disposal of 109,800 tons of C.W. ammunition. There are probably a further 100,000 tons which have not yet been examined and for which no programme has yet been drawn up. It is, however, proposed to load this at Nordafen."[176]

3 Sep 45   British Zone: Standing Committee on War
            Material (SCWM) meeting #19

The 19th meeting of the SCWM of the Control Commission for Germany (British Element) was held at 1500 hours, Monday, September 3, 1945 in the Chief of Staff's Conference Room, Koenigshof Hotel, Bad Oeynhausen. Brigadier K. R. Brazier-Creagh (Army Div.) was in the Chair. No U.S. representatives attended.

Priorities for dumping ammunition were again discussed:

"The Chairman stated that a reply had been received from the Secretary to the Continental Ammunition Dumping Committee to the effect that:

(a) his Committee (the SCWM) is responsible for co-ordinating demands of all three Services and for arranging a dumping programme,
(b) to date there had been no large-scale dumping by ship, but arrangements have now been made to clear 110,000 tons of CW ammunition by scuttling obsolete shipping in an approved area, and
(c) the programme for loading will shortly be sent out as an appendix to the minutes of the fourth meeting of the CADC."[177]

6 Sep 45   US: Minutes of Ammunition Committee Meeting

7 Sep 45   US: Declassified CONFIDENTIAL Memorandum for Record from Lt.Col. James N. Hinyard, Assistant Chief, Field Requirements Division, Chemical Warfare Service to all Theaters recommending the sea disposal of captured chemical munitions:

"At a conference attended by Colonel King, Colonel White, Lt.Col. Hinyard, Lt.Col. Totten, and Major Bradley, it was decided to make the following recommendation reference the disposition of unserviceable toxic munitions in the European Theatre of Operations: Due to hazards involved and excessive costs of handling toxics by any other method than dumping, recommend dumping at sea."[178]

11 Sept 45   2nd of 24 UK Atlantic Sea Disposal
              Operations (1 ship: Empire Simba)

17 Sep 45   British Zone: Standing Committee on War
              Material (SCWM) meeting #20

The 20th meeting of the SCWM of the Control
Commission for Germany (British Element) was held at
1500 hours, Monday, September 17, 1945 in the Chief of
Staff's Conference Room, Koenigshof Hotel, Bad
Oeynhausen. Brigadier K. R. Brazier-Creagh (Army Div.)
was in the Chair. No U.S. representatives attended.

Priorities for dumping ammunition were again addressed:
"The Chairman stated that as the three Services were
represented on the CADC, all arrangements for priorities
can be dealt with there and it is, therefore, unnecessary
for the Standing Committee on War Material to be
concerned in this matter unless facilities offered by the
Dumping Committee are not considered satisfactory, in
which case the Service representatives can raise the
matter at a meeting of this committee.

Captain R. H. Bevan, R.N. (Naval Division) pointed out
that the fiord outside Kiel Harbour was now to be used
for dumping ammunition and not Kiel Harbour itself.

The Committee agreed that arrangements for dumping
should not normally concern the Standing Committee on
War Material.[179]

17 Sep 45   British Zone: Continental Ammunition
              Dumping Committee (CADC) meeting #5

The fifth meeting of the Continental Ammunition
Dumping Committee (CADC) was held at HQ, BAOR on
September 17, 1945. Col C.H. Wooll, Q(M) Shipping, HQ.

BAOR, was in the Chair. Representing the U.S. Army were Lt.Col. L.C. Johnson and Major C.W. Schiller. The following information concerning the dumping of CW ex Germany was included in paragraph 3 of the minutes:

"The Chairman reported that progress is satisfactory at all ports and that the rate of loading is higher than forecast.

DD Tn (Ports) reported that there was a leakage of Phosgene in No. 1 hold of the ship loading at Nordhaven and it has been decided to stop further loading in that hold. Work is continuing throughout the rest of the ship.

Rep BAFO reported that authority has not yet been received to dump 40,000 tons of German aircraft bombs.

Lt. Col. J.A. Jaggers (Ord 4, BAOR) stated that they are not yet in a position to give accurate figures of the CW tonnage available over and above the first 110,000 but will do so as soon as possible."[180]

30 Sep 45   British Zone: CCG(BE) Progress Report #15

The 15th progress report of the Control Commission for Germany (British Element) covering the period 1st - 30th September 1945. The following information concerning CW disposal was excerpted from Appendix A, Naval Division Progress Report for Period 31st August - 30th September 1945; Appendix B, Army Division Progress Report; and Appendix D, Transportation Division Progress Report. Note the 10,000 ton discrepancy between the Navy and Army reports regarding the total amount of sea-disposed CW ammunition.

Naval Division: "The first convoy of five ships carrying approximately 17,000 tons of C.W. ammunition, which all belonged to the German Army, has sailed for the Skagerrak scuttling area. The loading of ships and hulks is now proceeding at the rate of about 2,000 tons a day."[181]

Army Division: "27,000 tons of CW ammunition has been dumped at sea out of a total so far uncovered of 120,000 tons."[182]

The estimate of 120,000 tons made in September 1945 was accurate; the BAOR Disarmament Progress Report of December 1947, which was the last BAOR report to include CW dumping information, confirmed that 119,910 tons of CW munitions had been disposed of from British control.

"Port Working - Emden. Total tonnage CW ammunition loaded: 9,521tons."[183]

1 Oct 45	3rd of 24 UK Atlantic Sea Disposal Operations (1 ship: Empire Cormorant)
4 Oct 45	British Zone: Chemical Weapons Convoy #1 (CW1)

The first British chemical weapons convoy (CW1) was scuttled. Five ships (the Triton, Patagonia, Louise Schroder, Pillau, and the Duborg), with a combined total of 17,100 tons of chemical munitions, were sent to the bottom.

4 Oct 45	British Zone: Standing Committee on War Material (SCWM) meeting #21

The 21st meeting of the Standing Committee on War

Material of the Control Commission for Germany (British Element) was held at 1500 hours, Thursday, October 4, 1945 in the Chief of Staff's Conference Room, Koenigshof Hotel, Bad Oeynhausen. Brigadier K. R. Brazier-Creagh (Army Div.) was in the Chair. No U.S. representatives were in attendance.

A progress report on dumping was offered: "Lt.Col. C. K. Simond (Army Division) stated that the programme for dumping was proceeding satisfactorily."[184]

5 Oct 45   British Zone: Continental Ammunition Dumping Committee (CADC) Meeting #6

The sixth meeting of the Continental Ammunition Dumping Committee was held at HQ, BAOR on October 5, 1945. Colonel C.H. Wooll, Q(M), Shipping BAOR, was in the Chair. No representatives from the U.S. Army were present. The minutes of this meeting are significant because they record the sailing and scuttling of the first CW convoy from North Germany:

"3. C.W. Ammunition
Commander A.A.E. Wheeler, BN C-in-C (G), stated that the first C.W. convoy from North Germany had been sailed and scuttled. This amounted to 17,000 tons. The names of the five ships were: Triton, Duborg, Pilau, Louise Schroder, and Patagonia. It was announced that to date a total of 45,463 tons of chemical warfare ammunition had been loaded."[185]

5 Oct 45   US: TOP SECRET telegram (now declassified) from the Joint Staff Mission in Washington to the Cabinet Offices.

This telegram is very significant in that it:

1) Communicates the policy decision that quadripartite agreement on the disposal of German CW stocks is unnecessary and

2) Reveals that the U.S. planned to fill any requirements for CW material from stocks available in the U.S. zone of occupation and authorized The War Office to similarly fill any U.K. requirements from captured stocks in the British zone.

"1. The United States Chiefs of Staff are informed that the Control Council (Germany) has agreed that the disposal of enemy war material will be the separate responsibility of each Zone Commander. In view of this fact, it is not considered useful to initiate action at this late date to secure quadripartite agreement to the policy on disposal of German chemical warfare stocks.

2. It is the intention of the United States Chiefs of Staff to fill as far as practicable U.S. requirements of chemical warfare material from stocks available in the U.S. zone of occupation. There is no objection to making available to the United Kingdom under arrangements agreeable to the Commanding General, U.S. Forces, European Theatre, any remaining stocks that may be desired for the British use, and it is suggested that such British requirements be made known directly to the Commanding General, U.S. Forces, European Theatre. Stocks of chemical warfare material remaining after the above requirements have been met will be destroyed."[186]

11 Oct 45   British Zone: RESTRICTED Letter (now
             declassified) from HQ, BAOR to The War
             Office, Subject: Disposal of Enemy CW
             Ammunition

"1. Up to 4 Oct 45, 45,463 tons of chemical ammunition
had been loaded on to hulks. The first convoy of five ships
which contained 17,000 tons was scuttled on 4 Oct 45.
The second convoy is due to sail on 11 Oct 45.

2. In addition to ammunition, certain stocks of solid
chemical warfare agents were uncovered in mines and at
Raubkammer are to be scuttled.

3. It is confirmed that No. 1 Porton Group had an unfilled
requirement for German Army ammunition. Accordingly,
the only natures which are not available for dumping are
Chemical Warfare Aircraft bombs. 5,000 rounds of
105mm Green Ring 3 shells are being held for the
LMAB."[187]

13 Oct 45   UK: TOP SECRET Letter (now declassified)
             from Major-General G. Brunskill, Director of
             Special Weapons and Vehicles, The War
             Office, to HQ, BAOR.

In this letter The War Office directed that the captured
German aircraft bombs marked Green Ring 3 and
charged with Tabun which were from an earlier
communication "frozen" (being retained for shipment to
the UK to augment British CW ammunition stocks) were
under Army control; the Royal Air Force was not to
destroy them.

"I am directed to refer to a telegram, JSM 73 of 5th October, 1945, from the Joint Staff Mission at Washington to the Cabinet Offices.

It is known that in fact you are already acting on a decision of the Control Council (Germany) to dispose of German Chemical Warfare material for which there is no requirement.

This decision would appear to give you discretion also to dispose of stocks of aircraft bombs, marked Green Ring 3 and charged with Tabun; which you have already been asked to hold for the present. I am to say that, for the present, this bomb should not be destroyed, for the view is held that there may be a requirement for Tabun or a similar agent in Army weapons, and until a better agent is available, the stocks of Tabun in these bombs provide our only supply of a valuable material.

I am to point out that it may be a matter for discussion whether these aircraft bombs are properly an Army or RAF responsibility; for the prima facie case that being in aircraft weapons, the Tabun is an RAF responsibility is countered by the knowledge that the material was charged into bombs as the most convenient means of storage and transport; and the material may be needed for Army purposes. The Air Ministry are, therefore, being requested to ensure that for the present at least no action by the RAF results in these bombs being destroyed." [188]

17 Oct 45   British Zone: Chemical Weapons Convoy #2 (CW2)

The second British chemical weapons convoy was sunk in the Skagerrak. Seven hulks were scuttled (the Balkan,

Drau, Edith Howaldt, Emmy Friedrich, Oderstrom, Olga Siemers, and the Trude Schunermann), with an estimated combined total of 32,215 tons of CW ammunition. The cumulative total for convoys 1 and 2 is 49,315 tons.

18 Oct 45   British Zone: Standing Committee on War Material (SCWM) meeting #22

The 22nd meeting of the Standing Committee on War Material of the Control Commission for Germany (British Element) was held at 1500 hours, Thursday, October 18, 1945 in the G(SD) Conference Room, No. 1, Porta Strasse, Bad Oeynhausen. Lt.Col. A. M. G. Dobson (Army Division) was in the Chair. No U.S. representatives were present.

30 Oct 45   4th of 24 UK Atlantic Sea Disposal Operations (1 ship: Wairuna)

31 Oct 45   British Zone: CCG(BE) Progress Report #16

The 16th progress report of the Control Commission for Germany (British Element) covered the period 1st - 31st October 1945. The following information concerning CW disposal was excerpted from Appendices A, B, and D: Naval, Army, and Transportation Division Progress Reports, respectively:

According to the Naval Division report, "a convoy of six ships (CW2) (actually seven ships), carrying approximately 24,000 tons of CW ammunition (actually 29,965 tons), which belonged to the German Army, has been sailed for the Skagerrak scuttling area, giving a total of 41,000 tons of CW ammunition dumped to date (actually 47,065 tons)."[189]

According to the Army Division report, however, "56,000 tons of German CW ammunition have been dumped at sea and approximately a further 13,000 tons have been disposed of by other means. This leaves 48,000 tons under control in the British Zone."[190] Note the 15,000 ton discrepancy between the Naval and Army Divisions' reports. Also according to the Army Division report "the ban on destruction of CW aircraft bombs, which had been frozen for possible use in the Far East, has now been lifted, with the exception of bombs charged 'TABUN' of which 8,000 tons are held. All other CW bombs are being included in the dumping programme."[191]

"Port Working - Emden. Total tonnage C.W. ammunition loaded: 11,716 tons. Port Working - Lubeck and Travemunde. Total tonnage exported (C.W. ammunition for dumping): 11,228 tons."[192]

Nov 45   The Allied Control Commission's Standing Committee on War Material submitted its thirteenth report entitled "Treatment of Surplus War Material", making specific recommendations on the disposal of chemical weapons. It stated that "dumping was the most practicable method of disposal" for most of these stocks and stated that Zonal Commanders should carry out the destruction or disposal of captured or surrendered German war material located in Germany as expeditiously as possible. The commander of each occupied zone was to furnish to the Allied Control Council a progress report on the disposal of such material. Each report was to itemise, by type, the total quantity of war material discovered, the total destroyed or otherwise disposed of, the cumulative amount of war material destroyed, and the balance of material for destruction.

15 Nov 45   British Zone: Standing Committee on War Material (SCWM) meeting #23

The 23rd meeting of the Standing Committee on War Material of the Control Commission for Germany (British Element) was held at 1500 hours, Thursday, November 15, 1945 in the Chief of Staff's Conference Room, Koenigshof Hotel, Bad Oeynhausen. Brigadier K. R. Brazier-Creagh (Army Division) was in the Chair. No U.S. representatives were in attendance.

17 Nov 45   British Zone: Chemical Weapons Convoy #3 (CW3)

The third British chemical weapons convoy was scuttled. Five ships (the Jantje Fritzen, Theda Fritzen, Sesostris, Tagila, and the Taurus), with a combined total of 15,416 tons of CM were sunk.

As an illustration of the inaccuracy and carelessness that characterized many of the dumping operations, while the Ministry of Defence claims the Sesostris was scuttled at 58°15'N 09°30'E, the Norwegian Defence Research Establishment (NDRE), during a 1989 investigation, found the Sesostris with a sidescan sonar (with an accuracy of ± 20 meters) at 58°18' 31.5"N 09°41' 05.7"E. Additionally, the MoD has no record of the Theda Fritzen, confusing it with the Jantje Fritzen. In an article published in the Hamburger Abendblott on August 19,

1970, Friedrich Passehl, captain on board the Taurus on its final expedition, confirmed there were five ships in the convoy.[193]

6 Dec 45   British Zone: Standing Committee on War Material (SCWM) meeting #24

The 24th meeting of the Standing Committee on War Material of the Control Commission for Germany (British Element) was held at 1500 hours, Thursday, December 6, 1945 in the War Establishments Committee Room, 2A Porta Strasse, Bad Oeynhausen. Lt.Col. J. L. Nicholls (Army Division, BAOR) was in the Chair. No U.S. representatives were present.

7 Dec 45   British Zone: Continental Ammunition Dumping Committee (CADC) Meeting #7

The seventh meeting of the Continental Ammunition Dumping Committee (CADC) was held at HQ, British Army of the Rhine (BAOR), on December 7, 1945. Lt.Col. J. H. Gordon (AQMG(M) Shipping, HQ, BAOR) was in the Chair. The following information was provided regarding the disposal of CW ammunition:

"7. PROGRESS OF DUMPING EX GERMANY

CW Amn

It was reported that tonnage loaded to 5 Dec was 75,336. Tonnage actually dumped 59,831. Three convoys had so far sailed and been scuttled.

BN C in C asked when fourth convoy would be ready for scuttling. It was estimated that provided smaller vessels are available for loading at Emden and Kiel, the fourth convoy will be ready by the end of December.

The Chairman stated that the reason for the slowdown in the program was:

(a) limitation of rail lift to 1,000 tons per day.

(b) high proportion of light bombs and lack of bottom stowage cargo.

Hulks still available for loading total approximately 50,000 tons.

CW Tonnage on US Account

It was agreed by US Representative OCOT (Major F.W. Harvey) that full details of the breakdown and location of the 24,000 tons on US account would be supplied together with a percentage of bottom stowage, i.e. heavy cased ammunition. He thought the majority of the ammunition was heavy cased.

8. AVAILABILITY OF HULKS FOR LOADING CW EX GERMANY

Q(M) 30 Corps asked for arrangements to be made for the provision of smaller hulks for loading at Emden. It was reported that the next vessel available for loading was the Karl Leonhardt of 8,000/10,000 tons capacity. This vessel would need at least 2,000 tons of bottom stowage cargo before a start could be made on loading CW bombs. It was agreed that MWT (Rotterdam) be asked as a matter of urgency to supply hulks of 1,500/2,000 tons capacity at Emden and Kiel for loading the remainder of the programmed tonnage.

9. AVAILABILITY OF HULKS

BN C in C stated that some of the hulks listed for CW dumping on the original programme may have since been allocated to the Russians under the Quadripartite agreement. To offset this, other hulks may be available ex Bremen enclave."

11. ACTION OF DUMPING COMPLETION
G(SD) HQ, BAOR (Major D.H. Stuart-Brown) asked, in view of the urgency to complete the dumping programme, an estimated completion date could be given. It was agreed that in view of the difficulties likely to be encountered during the winter months with storms and weather conditions, that no firm date could be given. The Chairman, however, stressed that the maximum effort must be made by all concerned both by careful programming of movement and loading, to ensure that at least 1,000 tons per day overall is moved and dumped.

15. MOVEMENT OF CW BY RAIL IN CLOSED VANS
Q(M) 30 Corps stated that a proportion of CW had been loaded in closed vans. This raised difficulties of decontamination and discharge. It was agreed by Q(M)5 HQ, BAOR (Major J.W. Walford) that in the future, open wagons only will be provided for CW loading."[194]

20 Dec 45   British Zone: Standing Committee on War Material (SCWM) meeting #25

The 25th meeting of the Standing Committee on War Material of the Control Commission for Germany (British Element) was held at 1500 hours, Thursday, December 20, 1945 in the Chief of Staff's Conference Room, Koenigshof Hotel, Bad Oeynhausen. Brigadier G. H. Baker, B.G.S. (SD), HQ, BAOR, was in the Chair. No U.S. representatives attended.

27 Dec 45   American Zone: Brigadier General Henry C.

> Wolfe, U.S. Army, Acting Director, Plans and Operations, ASF, recommended that theatre commanders dispose of ammunition which will become hazardous due to storage conditions by the time shipping becomes available to move such ammunition.[195]

30 Dec 45   5th of 24 UK Atlantic Sea Disposal Operations (1 ship: Botlea)

1945-1946   The U.S. disposed of "unspecified amounts" of CW in the Mediterranean Sea off the west coast of Italy near Naples and off the east coast near both Bari and Brindisi.

4 Jan 46   U.S. War Department: Message CM-OUT-91306, 0106 hours, January 4, 1946 from Captain Stanton, Requirements Branch, Requirements Division (Ordnance-Chemical Warfare Service), Headquarters ASF, to Commanding Generals in various theatres authorized the disposal of hazardous ammunition by dumping at sea or demolition:

"Paragraph 2 of letter, AGMP-M 400.7 (17 September 1945) SPOPP, subject "Methods of Disposing of Surplus Ammunition", dated 18 September 1945, authorizes disposal of hazardous ammunition by dumping at sea or demolition in accordance with existing regulations. This paragraph may be interpreted to include ammunition which will become hazardous due to storage conditions by the time shipping becomes available to move such ammunition."[196]

10 Jan 46   British Zone: Standing Committee on War
              Material (SCWM) meeting #26

The 26th meeting of the Standing Committee on War
Material of the Control Commission for Germany (British
Element) was held at 1500 hours, Thursday, January 10,
1946 in the Chief of Staff's Conference Room, Koenigshof
Hotel, Bad Oeynhausen. Brigadier K. R. Brazier-Creagh
(now B.G.S. - G (SD), vice Army Division) was in the
Chair. No U.S. representatives were in attendance.

19 Jan 46   American Zone: Message Number S-38304
              (CM-IN 4442), 1610 hours, January 19,
              1946, from the Commanding General,
              USFRT, Main, Frankfurt, Germany
              (McNarney) to the U.S. War Department.

"British Army of the Rhine (BAOR) has indicated that
they would assist United States Army in disposing of
captured chemical ammunition located in the United
States Zone of Germany east of Rhine by calling forward
for sea dumping an unstated quantity on a daily
movement basis when their facilities will permit. They
have requested information as to the quantity and type
by location of chemical ammunition available for calling
forward.

It is estimated that there are approximately 90,000 tons
of captured chemical ammunition located in the United
States Zone of Germany east of Rhine. This material
constitutes a continuous threat and hazard because of a
lack of personnel to adequately guard and maintain it. It
is urged that authority be given to advise the British as
to types and quantity by location as they have requested
and that this authority provide for the shipment for sea

dumping of whatever quantities the British are able to call forward."[197]

21 Jan 46   British Zone: Continental Ammunition Dumping Committee special meeting

A special meeting of the Continental Ammunition Dumping Sub-Committee was called on January 21, 1946 at HQ, BAOR to discuss the dumping through Belgian ports of approximately 10,000 tons of British high explosive (HE) ammunition located in Belgium. LtCol J. H. Gordon, Assistant Quartermaster General (M), Shipping, HQ, BAOR, was in the Chair.

22 Jan 46   U.S. War Department: Radio CM-OUT 94065 advised the Commanding General, USFRT, Main, Frankfurt, Germany, that the authorities requested in message S-38304 (CM-IN 44420, January 19, 1946) were granted by the U.S. Department of State.[198]

23 Jan 46   British Zone: SECRET Letter (now declassified) from HQ, BAOR to The War Office, Subject: Disposal of Enemy CW Stocks.

"Up to 15 January 1946, approximately 80,000 tons of enemy CW ammunition had been moved to hulks for scuttling. Three convoys totaling 64,731 tons have been scuttled and a fourth convoy will sail shortly.

28,478 tons available for dumping are now held in Raubkammer, Munsterlager (Depot 24), and Orrel (Depot 33). Eventually, all stocks of "frozen" material, i.e. 8,303 tons of Green Ring 3 aircraft bombs, 1,050 250kg White

Ring bombs, and 40 tons of bulk CN will be held at Orrel pending further instructions.

All bulk stocks of liquid chemical warfare agents amounting to some 4,000 tons have been destroyed by burning. Destruction of smaller quantities of chemical warfare material at Raubkammer continues.

<u>In addition to German CW stocks, 24,000 tons of U.S. Army chemical warfare ammunition are beginning to move to Kiel for inclusion in the present scuttling programme."</u>[199]

24 Jan 46   British Zone: Standing Committee on War Material (SCWM) meeting #27

The 27th and penultimate meeting of the Standing Committee on War Material of the Control Commission for Germany (British Element) was held at 1500 hours, Thursday, January 24, 1946 in the War Establishments Committee Room, 2A Porta Strasse, Bad Oeynhausen. Brigadier K. R. Brazier-Creagh (B.G.S. - G (SD), HQ, BAOR) was in the Chair. No U.S. representatives were present. The Committee, inter alia, discussed the relationship between the SCWM and the Standing Committee on Demilitarization:

"The Chairman made a short statement on the functions of the Standing Committee on Demilitarisation with particular reference to its effect on the present committee. The Standing Committee on Demilitarisation had had its first two meetings. It had been formed as a link between Army Division in Berlin and HQ, BAOR. The functions of the Committee are:

(a) to advise the Chief of the Army Division on matters of policy within the scope of that Division, particularly as concerns the practical implementation of policy and its effect on the operation of HQ, BAOR;

(b) to act as the channel of communication between the Army Division and HQ, BAOR; and

(c) to ensure by all necessary means that policy directions received from the Army Division are fully executed."[200]

Feb 46   American Zone: The 18th Chemical Maintenance Company assumed responsibility for Grafenwöhr Chemical Depot. After the arrival of the 18th Chemical Maintenance Company, a work program was inaugurated including restacking, collection of scattered items, demilitarization, decontamination, destruction, and boxing for later shipment.

Feb 46   British Zone: HQ, BAOR issued detailed instructions concerning the disposal of German ammunition. CW munition stocks were to be moved by rail to ports in the 8 Corps area, loaded into hulks, towed out to sea under arrangements made by the Royal Navy, and finally scuttled in the Skagerrak. These instructions confirm the approved method of disposal was dumping in seawater at a depth of approximately 300 fathoms (549 meters).

21 Feb 46   American Zone: Theatre Headquarters authorized the major commands to start the destruction of all enemy ground forces

material of a warlike nature, with the proviso that the destruction would not be undertaken until all Theatre requirements for such material had been determined by the Theatre Chiefs of Technical Services.[201]

27 Feb 46   French Zone: SECRET letter (now declassified) from HQ, BAOR to The War Office, Subject: CW Ammunition Held by the French

"1. The following information concerning stocks of German CW ammunition held by the French has been communicated to this Headquarters by Lt. Col. D'Anselme, Director Generale du Controle da Desarmament, Commandement en Chef Francais en Allemagne, Baden-Baden.

2. A total quantity of 8,000 tons of CW ammunition is held by the French at the Urlau Depot (near to Constance), this being made up as follows:

(a) 2,000 tons, shell charged "DM",
(b) 3,000 tons, shell charged "H", and
(c) 3,000 tons, shell charged TABUN.

3. Items in para 2(a) and (b) are being destroyed by the French, while assistance is being sought from this HQ (HQ, BAOR) with regard to the destruction of the TABUN (sub-para 2(c)) by deep sea dumping.[202]

16 Mar 46   British Zone: Chemical Weapons Convoy #4
(CW4)

Four months had now passed since CW3. The fourth
British chemical weapons convoy was scuttled. Four ships
(the Falkenfels, Fechenheim, Hugo Oldendorf, and the
Karl Leonhardt) with an estimated combined total of
18,000 tons of CM (not including the Fechenheim and the
Hugo Oldendorf, for which cargo data is unavailable)
were sunk.

18 Mar 46   British Zone: Continental Ammunition
Dumping Committee (CADC) Meeting #8

The 8th meeting of the CADC was held at HQ, BAOR, on
March 18, 1946. Lt.Col. J. H. Gordon, Assistant
Quartermaster General (M), Shipping, HQ, BAOR, was in
the Chair.

18 Mar 46   American Zone: The mustard (H) bomb-
filling program started at St. Georgen
Chemical Depot.

21 Mar 46   British Zone: Standing Committee on War
Material (SCWM) meeting #28

The 28th and final meeting of the Standing Committee on
War Material of the Control Commission for Germany
(British Element) was held at 1500 hours, Thursday,
March 21, 1946 in the Chief of Staff's Conference Room,
Koenigshof Hotel, Bad Oeynhausen. Brigadier K. R.
Brazier-Creagh (B.G.S.- G (SD), HQ, BAOR) was in the
Chair. As had been the case since meeting #12 on April 4,
1945, no U.S. representatives were present.

"The Chairman informed the meeting that this was the
last meeting of the Committee, whose functions would be

taken over by the Inter-Services Executive Committee."[203]

18 Apr 46   American Zone: The mustard (H) bomb-filling program was completed at St. Georgen Chemical Depot. 9,885 aerial bomb casings were filled with liquid mustard between 18 March and 18 April 1946. The bombs were then forwarded to the port of Bremen and shipped to Edgewood Arsenal, Maryland in May 1946.

26 Apr 46   Berlin: In order to provide for a uniform method of reporting the materials found in the four zones, the Allied Control Council issued Directive No, 28, which provided that destruction or disposal of captured or surrendered German war material should be completed as expeditiously as possible. In addition, the Commander of each zone was required to furnish to the Allied Control Authority a semi-annual progress report, as of 1 June and 1 December, on the disposal of German war materials. The report was to include the following information:

a. Total war material found to date.
b. Account of war material destroyed or otherwise disposed of during the period (normally six months) covered by the report.
c. Accumulated amounts of war material destroyed or otherwise disposed of to date.
d. Remainder of was material for destruction or disposal.[204]

27 Apr 46   American Zone: A mustard-burning project

started at St. Georgen Chemical Depot.

Apr 46   American Zone: In April 1946, ten hulks were transferred to the War Department from the War Shipping Administration and were made available to the Theatre for filling with ammunition and scuttling. The work of dumping, scuttling, and destruction was accelerated. Plans called for the disposal of all captured enemy ammunition in one year, but required an increase in personnel, both Army and civilian.[205]

1 May 46   Shipments of CW to the port of Bremerhaven for scuttling and to Antwerp for shipment back to the U.S. had started.

28 May 46   American Zone: Representatives of the Office of Military Government for Germany (U.S.) (OMGUS) and Theatre Headquarters met with German officials to plan the turnover of all remaining stocks of captured enemy ammunition to OMGUS for demilitarization and salvage of component parts for use in the German economy. The plan provided for the transfer of approximately 250,000 tons of non-toxics and 70,000 tons of toxics for breakdown into component raw materials for use in the German economy in connection with the agriculture and manufacturing program.[206]

Jun 46   American Zone: By June 1946, all U.S. requirements for CW from captured German stocks had been shipped, destruction of CW at depots in situ was underway, and initial sinkings of loaded hulks had started.[207]

24 Jun 46   American Zone: The plan conceived on May 28th was approved. All scuttling operations and in situ destruction at depots was halted on order from USFET. Final destruction responsibility was transferred from USFET/Continental Base Section to OMGUS.

1 Jul 46   American Zone: "Davey Jones Locker" Chemical Weapons Convoy #1 (DJL1)

This was the first American chemical weapons convoy of "Operation Davey Jones Locker". [Note the first British chemical weapons convoy (CW1) was scuttled 9 months earlier, on October 4, 1945. The first American sea disposal operations actually occurred in December 1945.] Two ships, the Sperrbrecher and the T-65, were scuttled in the Skagerrak (at 58°14' N 9°15' E and 58°17'9" N 9°37'1" E, respectively) with a combined total of 3,220 tons (2,875 long tons) of munitions containing tabun (GA), mustard (H), phosgene (CG), and chloracetophenone (CNB). They were the first of eleven ships sunk in this operation. The Sperrbrecher and T-65 were towed out to sea by tugs. It is not known how the Sperrbrecher was sent to the bottom. The seacocks were opened and shells were fired into the hull of the T-65 at the water line in the bow and stern where no toxics were stored. According to official records, both are at 650 meters.

2 Jul 46   American Zone: "Davey Jones Locker" Chemical Weapons Convoy #1 (DJL1)

The U.J.-305, a German trawler, was scuttled in the Skagerrak (at 58°16'4" N 9°29' E) containing 751.5 tons (671 long tons) of CW. The seacocks were opened and shells were fired into the ship's hull at the water line in the bow and stern where no toxics were stored. According to official records, the U.J.-305 is at 650 meters.

2 Jul 46   Norway: The Norwegian newspaper Fædrelandsvennen published an article entitled "48,000 tonn giftgass skal senkes I Skagerakk" (48,000 Tons of Poisonous Gas Scuttled in the Skagerrak) citing an interview with U.S. Marine Commander Joaquim Flood.

Commander Flood, according to the article, led a convoy of three German ships filled with 3,800 tons of tabun and mustard gas. He explained that the bottom valves were opened, and water flooded to cover the 100,000 gas containers. In addition, the American ships opened fire to let the ships sink faster.[208] This article was obviously referring to U.S. convoys DJL1 and DJL2 of 1 and 2 July, respectively. The actual amount of CW scuttled in these three ships was 3,972 tons.

3 Jul 46   American Zone: The mustard burning project at St. Georgen Chemical Depot was completed.

4,138 metric tons of captured German liquid mustard and 745 metric tons of Clark I and Clark II were disposed of by means of burning. All work was carried out by German civilian laborers under the close supervision of American officers and enlisted men. Activities were coordinated with local civilian authorities. Mustard was burned only on those days when wind would carry the smoke into the previously designated area and only when the wind was blowing not less than three miles or more than ten miles per hour. Winds and weather conditions were constantly checked and burning was halted whenever there was sufficient change in wind direction or velocity to carry the smoke outside of the previously designated limits. Considerable concern on the part of the German populace was felt at the beginning of the burning program which subsided as time went on and no evidence of any damage could be detected. As far as could be determined, there were no serious or lasting injuries to life or property.[209]

13 Jul 46   British Zone: Chemical Weapons Convoy #5 (CW5)

The fifth British chemical weapons convoy was scuttled. Two ships (the Freiburg and the Gertrud Fritzen) were sunk with a combined total of 11,000 tons of CM.

14 Jul 46   American Zone: "Davey Jones Locker" Chemical Weapons Convoy #2 (DJL2)

The Alco Banner, a hog-islander, was scuttled in the Skagerrak (at 58°18'7" N 9°36'5" E) containing 3,096.8 tons (2,765 long tons) of CW. It took 45 minutes for the Alco Banner to sink and was sent to the bottom by shellfire. It sunk by plunging her bow in and going under without capsizing. According to official records, the Alco Banner is at 650 meters.

Jul-Oct 46   The U.S. disposed of 1,700 tons of mustard and lewisite-filled bombs off the coast of St. Raphael, France.

25 Aug 46   6th of 24 UK Atlantic Sea Disposal Operations (1 ship: Empire Peacock)

29 Aug 46   British Zone: CCG(BE) Demilitarisation Committee meeting #1.

The first meeting of the Demilitarisation Committee of the Control Commission for Germany (British Element) commenced at 1500 hours, Thursday, August 29, 1946 in the Tax House, Lübbecke. A total of 32 meetings were held between Aug 29, 1946 and December 8, 1949. Major General P. M. Balfour, DCOS(X) was in the Chair. Brigadier K. R. Brazier-Creagh represented HQ, BAOR; Captain P.L. Saumarez, R.N. represented Naval Division; and Lt.Col. J. L. Nichols represented Army Division. The Committee invited HQ, BAOR to make a verbal statement at each meeting on the progress made in dumping enemy ammunition.[210]

30 Aug 46   American Zone: "Davey Jones Locker" Chemical Weapons Convoy #3 (DJL3)

The James Otis, an American liberty ship, was scuttled in the Skagerrak (at 58°16' N 9°32' E) containing 4,091.4 tons (3,653 long tons) of CW. This was the last American chemical weapons convoy in 1946. It is not known how the James Otis was sent to the bottom.

Sep 46   American Zone: At Grafenwöhr Chemical Depot the consolidation of like types of ammunition was completed and inventories were carefully checked for the final step in

turning over these stocks to the Military
Government for conversion into peacetime
products.

3 Sep 46   7th of 24 UK Atlantic Sea Disposal
Operations (1 ship: Empire Nutfield)

8 Sep 46   British Zone: Chemical Weapons Convoy #6
(CW6)

The sixth British chemical weapons convoy was sunk.
Two ships (the Deutchland and the Rhon), with a
combined total of 2,121 tons of CM were sunk in the
Skagerrak. Commencing with CW6 on September 8,
1946, the British CW sea-dumping program was mounted
solely through the port of Emden.[211]

26 Sep 46   British Zone: CCG(BE) Demilitarisation
Committee meeting #2

The second meeting of the Demilitarisation Committee of
the Control Commission for Germany (British Element)
commenced at 1500 hours, Thursday, September 26, 1946
in the Tax House, Lübbecke. Major General W. H. A.
Bishop, DCOS(X) was in the Chair. Lt. Col. A. C. Hordern
represented HQ, BAOR; Brigadier C. J. G. Dalton
represented Army Division; and Captain F. S. W. de
Winton, R.N. represented Naval Division. HQ, BAOR
reported the dumping programme was going well. "8,000
tons of chemical warfare ammunition are still to be
dumped but enough hulks are available and dumping
should be completed before the end of the year." [212]

Oct 46   All remaining toxics in the U.S. Zone began
to be consolidated at St. Georgen.

1 Oct 46   8th of 24 UK Atlantic Sea Disposal
            Operations (1 ship: Kindersley)

12 Oct 46   British Zone: Chemical Weapons Convoy #7
            (CW7)

The seventh British chemical weapons convoy was
scuttled. Three ships (the Eider, Empire Severn, and the
Ludwigshaven), with a combined total of 11,421 tons of
CW munitions, were sunk.

31 Oct 46   British Zone: CCG(BE) Demilitarisation
            Committee meeting #3

The third meeting of the Demilitarisation Committee of
the Control Commission for Germany (British Element)
commenced at 1500 hours, Thursday, October 31, 1946 in
the Tax House, Lübbecke. Major General W. H. A.
Bishop, DCOS(X) was in the Chair. Major H. M. Lang
represented HQ, BAOR; Lt.Col. J. L. Nichols represented
Army Division; and Captain F. S. W. de Winton, R.N.
represented Naval Division. A verbal statement was
made by HQ, BAOR on the dumping of CW ammunition.
"When the present hulk is completed, in ten to fourteen
days time, 4,500 tons will remain to be dumped."[213]

2 Nov 46   9th of 24 UK Atlantic Sea Disposal
            Operations (1 ship: Empire Woodlark)

11 Nov 46   10th of 24 UK Atlantic Sea Disposal
            Operations (1 ship: Lanark)

15 Nov 46   British Zone: CCG(BE) Demilitarisation
            Committee meeting #4 (Special Meeting)

A special meeting of the Demilitarisation Committee of
the Control Commission for Germany (British Element)

commenced at 1500 hours, Friday, November 15, 1946 in the Tax House, Lübbecke. Major General W. H. A. Bishop, DCOS(X) was in the Chair. Lt.Col. A. C. Hordern represented HQ, BAOR and Captain F. S. W. de Winton, R.N. represented Naval Division. Army Division was not represented. No report on the disposition of CW was made.

12 Dec 46   British Zone: CCG(BE) Demilitarisation Committee meeting #5

The fifth meeting of the Demilitarisation Committee of the Control Commission for Germany (British Element) was held on Thursday, December 12, 1946 in the Tax House, Lübbecke. Major General W. H. A. Bishop, DCOS(X) was in the Chair. Lt.Col. A. C. Hordern represented HQ, BAOR; Captain F. S. W. de Winton, R.N. represented Naval Division; and Brigadier V. Paley represented the new Combined Services Division (Army Branch). This is the last Demilitarisation Committee meeting at which a report was given concerning the dumping of enemy ammunition. "Between 12,000 and 13,000 tons of CW ammunition remain for dumping. The bulk of this should be complete by the end of January 1947, and any small quantity remaining will be disposed of steadily."[214]

Dec 46   UK: Report by the Prime Minister's Commission on Demilitarisation in the British Zone of Germany (SECRET report now declassified)

Major-General R. G. Stone, President of the Prime Minister's Commission on Demilitarisation in the British Zone of Germany presented the following information concerning the disposition of enemy CW in his "Report of Progress of Disarmament and Demilitarisation in British Zone in Germany."

War Material - Situation on 1st May 1946:
"Chemical warfare ammunition 74% disposed. Target date for completion: December 1946."[215]

Chemical Warfare Ammunition:
"Ammunition of this type has usually to be dumped at sea. There is at present a serious shortage of hulks for this purpose which, if not remedied, will cause a slowing down of the rate of disposal."

"There is in the British zone about 8,000 tons of the German secret gas Tabun. This is at present 'frozen' by instructions from London. A further 10,000 tons may have to be accepted from the U.S. Zone."[216]

Disposal of Tabun:
"It will be difficult to justify the holding of Tabun in the British Zone and a policy decision on the disposal of this gas is required from London."[217]

Chemical Warfare Equipment:
Bulk Gases:
Total found to date: 4,172 metric tons
Destroyed or disposed of to date: 3,010 metric tons
Balance for disposal: 1,162 metric tons

Chemical Warfare Ammunition:
Total found to date: 117,913
Destroyed or disposed of to date: 87,638
Balance for disposal: 30,275 [218]

21 Dec 46   British Zone: Chemical Weapons Convoy #8 (CW8)

The eighth British chemical weapons convoy, comprised of only two ships (the Monte Pascoal and the T-(Z) 63), each with an unknown cargo, was sunk.

Note: Because of the difficulties all parties were encountering in completing their respective disposal programs, the initial date agreed by the Allied Control Council for the destruction of all chemical weapons, the end of 1946, was subsequently extended by general agreement to the end of 1947.[219]

1946-1978   Russian loose-dump disposal operations conducted in the Baltic Sea

23 Jan 47   British Zone: CCG(BE) Demilitarisation Committee meeting #6

The sixth meeting of the Demilitarisation Committee of the Control Commission for Germany (British Element) was held on Thursday, January 23, 1947 in the Tax House, Lübbecke. Major General W. H. A. Bishop, DCOS(X) was in the Chair. Lt.Col. A. C. Hordern represented HQ, BAOR; Commander G. H. Anderson represented Naval Division; and Lt.Col. J. L. Nichols represented the Combined Services Division (Army Branch). No report on the disposition of CW was made.

Jan-Feb 47   American Zone: All toxic munitions remaining at Grafenwöhr Chemical Depot were shipped to the St. Georgen Chemical Depot.

5 Feb 47   11th of 24 UK Atlantic Sea Disposal Operations (1 ship: Dora Oldendorf)

27 Feb 47   British Zone: CCG(BE) Demilitarisation
             Committee meeting #7

The seventh meeting of the Demilitarisation Committee
of the Control Commission for Germany (British
Element) was held on Thursday, February 27, 1947 in the
Tax House, Lübbecke. Major-General W. H. A. Bishop,
DCOS(X) was in the Chair. Lt.Col. G. P. Warden
represented HQ, BAOR; Captain F. S. W. de Winton, R.N.
represented Naval Division; and Brigadier V. Paley
represented the Combined Services Division (Army
Branch). No report on the disposition of CW was made.
"The Chairman proposed that the name of the Committee
should be changed to 'Zonal Demilitarisation Committee'.
The Committee agreed to this change."[220]

15 Mar 47   American Zone: The Grafenwöhr Chemical
             Depot was closed after being under
             American control for 695 days (April 19,
             1945 - March 15, 1947) and the 18th
             Chemical Maintenance Company was
             deactivated.

15 Mar 47   American Zone: The Wildflecken Chemical
             Depot was closed after being under
             American control for 708 days (April 6, 1945
             - March 15, 1947) and the 13th Chemical
             Maintenance Company was deactivated.

27 Mar 47   British Zone: CCG(BE) Zonal
             Demilitarisation Committee meeting #8

The eighth meeting of the Zonal Demilitarisation
Committee of the Control Commission for Germany
(British Element) was held on Thursday, March 27, 1947

in the Tax House, Lübbecke. Brigadier D. Meynell, ACOS(B) was in the Chair. Lt.Col. G. P. Warden represented HQ, BAOR; Lt.Cdr. B. C. Ward, R.N. represented Naval Division; and Lt.Col. J. L. Nichols represented the Combined Services Division (Army Branch). No report on the disposition of CW was made.

1 Apr 47 American Zone: All remaining toxics in the American Zone were consolidated at St. Georgen Depot by April 1, 1947 and a detailed inventory was conducted on 15 April. Operation "Davey Jones Locker", which had been halted since convoy DJL4 was scuttled on August 30, 1946, was about to resume.

10 Apr 47 American Zone: The Frankenberg Chemical Depot was closed after being under American control for 742 days (March 29, 1945 - April 10, 1947) and the 227th Chemical Base Depot Company was deactivated.

10 Apr 47 American Zone: The Schierling Chemical Depot was closed after being under American control for 711 days (April 29, 1945 - April 10, 1947) and the 140th Chemical General Service Company was deactivated.

15 Apr 47 American Zone: All remaining toxics in the American Zone were consolidated at St. Georgen Depot by April 1, 1947 and a detailed inventory was conducted on 15 April.

Apr 47   British Zone: CW9. Having been loaded with an unknown amount of CW by February 20, 1947, the penultimate chemical weapons convoy, CW9, comprised of the single hulk "Dessau", was scuttled.

9 May 47   British Zone: CCG(BE) Zonal Demilitarisation Committee meeting #9

The ninth meeting of the Zonal Demilitarisation Committee of the Control Commission for Germany (British Element) was held on Friday, May 9, 1947 in the Tax House, Lübbecke. Major-General W. H. A. Bishop, DCOS(X) was in the Chair. Lt.Col. G. P. Warden represented HQ, BAOR; Lt.Cdr. B. C. Ward, R.N. represented Naval Division; and Lt.Col. J. L. Nichols represented the Combined Services Division (Army Branch). No report on the disposition of CW was made.

12 May 47   American Zone: Rail shipments from St. Georgen to the port at Bremerhaven, Germany began.

17 May 47   Convoy CW9 (9th UK Skagerrak CW Convoy, 1 ship)

1 Jun 47   American Zone: St. Georgen Chemical Depot was closed after being under American control for 759 days (May 3, 1945 - June 1, 1947).

6 Jun 47   American Zone: "Davey Jones Locker" Chemical Weapons Convoy #4 (DJL4)

The James Sewell, an American liberty ship, was scuttled in the Skagerrak (at 58°15'02" N 09°30'06" E) containing 4,480 tons (4,000 long tons) of CW. Dumping had resumed after a halt of 280 days (over 9 months). The previous convoy was scuttled on 30 Aug 46.

6 Jun 47   British Zone: CW10

In spite of difficulties in obtaining sufficient, suitable hulks, and with the severe weather of the winter of 1946-47 when some hulks were ice-bound, the last CW-loaded hulk, the Schwabenland, was scuttled in the Skagerrak on June 6, 1947, thus ending the ten-convoy, 31-ship, 610-day, sea-dumping program (October 4, 1945 - June 6, 1947).

19 Jun 47   British Zone: CCG(BE) Zonal
              Demilitarisation Committee meeting #10

The tenth meeting of the Zonal Demilitarisation Committee of the Control Commission for Germany (British Element) was held at 1430 hours, Thursday, June 19, 1947 in the Tax House, Lübbecke. Brigadier D. Meynell, ACOS(B) was in the Chair. Lt.Col. G. P. Warden represented HQ, BAOR and Brigadier V. Paley represented the Combined Services Division (Army Branch). There was no representative from the Naval Division and no report was made on the disposition of CW.

20 Jun 47   American Zone: "Davey Jones Locker"
              Chemical Weapons Convoy #5 (DJL5)

The James Harrod, an American liberty ship, was scuttled in the Skagerrak (at 58°16' N 09°33' E) containing 3,360 tons (3,000 long tons) of CW.

30 Jun 47   American Zone: "Davey Jones Locker"
              Chemical Weapons Convoy #6 (DJL6)

The George Hawley, an American liberty ship, was scuttled in the Skagerrak (at 58°18'05" N 09°38' E) containing 1,120 tons (1,000 long tons) of CW.

18 Jul 47   American Zone: "Davey Jones Locker" Chemical Weapons Convoy #7 (DJL7)

The Nesbitt, an American liberty ship, was scuttled in the Skagerrak (at 58°15' N 09°30' E) containing 6,720 tons (6,000 long tons) of CW. This was the last of 9 ships scuttled in the Skagerrak by the Americans. The remaining two ships of Operation Davey Jones Locker (comprising convoy DJL9) were scuttled in the North Sea over a year later.

24 Jul 47   British Zone: CCG(BE) Zonal Demilitarisation Committee meeting #11

The eleventh meeting of the Zonal Demilitarisation Committee of the Control Commission for Germany (British Element) was held on Thursday, July 24, 1947 in the Tax House, Lübbecke. Brigadier D. Meynell, ACOS(B) was in the Chair. Lt.Col. G. P. Warden represented HQ, BAOR and Major J. P. C. Bindloss represented the Combined Services Division (Army Branch). There was no representative from the Naval Division and no report was made on the disposition of CW.

27 Jul 47   12th of 24 UK Atlantic Sea Disposal Operations (1 ship: Empire Lark)

9 Aug 47   13th of 24 UK Atlantic Sea Disposal Operations (1 ship: Leighton)

8 Sep 47   14th of 24 UK Atlantic Sea Disposal Operations (1 ship: Thorpe Bay)

12 Sep 47   British Zone: CCG(BE) Zonal
              Demilitarisation Committee meeting #12

The twelfth meeting of the Zonal Demilitarisation
Committee of the Control Commission for Germany
(British Element) was held at 1430 hours, Friday,
September 12, 1947 in the Tax House, Lübbecke. Major-
General W. H. A. Bishop, DCOS(X) was in the Chair. For
the first time, Brigadier K. R. Brazier-Creagh
represented HQ, BAOR; Lt.Cdr. B. C. Ward, R.N.
represented Naval Division; and Brigadier V. Paley
represented the Combined Services Division (Army
Branch). No report was made on the disposition of CW.

17 Oct 47   British Zone: CCG(BE) Zonal
              Demilitarisation Committee meeting #13

The 13th meeting of the Zonal Demilitarisation
Committee of the Control Commission for Germany
(British Element) was held at 1430 hours, Thursday,
October 17, 1947 in the Tax House, Lübbecke. Major-
General W. H. A. Bishop, DCOS(X) was in the Chair.
Brigadier K. R. Brazier-Creagh represented HQ, BAOR;
Lt.Cdr. B. C. Ward, R.N. represented Naval Division; and
Brigadier V. Paley represented the Combined Services
Division (Army Branch). No report was made on the
disposition of CW.

3 Nov 47   15th of 24 UK Atlantic Sea Disposal
              Operations (1 ship: Margo)

13 Nov 47   British Zone: CCG(BE) Zonal
              Demilitarisation Committee meeting #14

The 14th meeting of the Zonal Demilitarisation
Committee of the Control Commission for Germany

(British Element) was held at 1430 hours, Thursday, November 13, 1947 in the Tax House, Lübbecke. Brigadier J. N. D. Tyler, ACOS(Exec)B was in the Chair. Brigadier K. R. Brazier-Creagh represented HQ, BAOR and Lt.Cdr. B. C. Ward, R.N. represented Naval Division. The Combined Services Division (Army Branch) was not represented and no report was made on the disposition of CW.

18 Dec 47   British Zone: CCG(BE) Zonal
             Demilitarisation Committee meeting #15

The 15th meeting of the Zonal Demilitarisation Committee of the Control Commission for Germany (British Element) was held at 1430 hours, Thursday, December 18, 1947 in the Tax House, Lübbecke. Brigadier J. N. D. Tyler, ACOS(Exec)B was in the Chair. Brigadier K. R. Brazier-Creagh represented HQ, BAOR and Major W. J. Bennett represented the Combined Services Division (Army Branch). The Naval Division was not represented and no report was made on the disposition of CW.

Dec 47   British Zone: By December 1947, the small
             quantity of CW munitions remaining after
             the scuttling of CW Convoy #10 (CW10), the
             sea-dumping program's final convoy, was
             disposed of by burning.[221]

Dec 47   British Zone: The last BAOR Disarmament
             Progress Report to mention CW stocks is
             that of December 1947 which states that a
             total of 119,910 tons of CW munitions had
             been disposed of from British control.[222]

According to the British Ministry of Defence, the surviving records of the sea-dumping programme indicate that at least 127,000 tons of CW munitions were disposed of by this means. This figure includes 3,000 tons from the French Zone and up to 10,000 tons from the American Zone, these latter quantities being sea-dumped by Britain on their Allies' behalf. 10,000 tons of CW munitions recovered in the British Zone were shipped to the U.K. for research purposes under the code name "Op Dismal".[223]

Research conducted by the German Federal Ministry of Transport and the Helsinki Commission's Ad Hoc Working Group on Dumped Chemical Munition, however, indicates that 122,508 tons of CW munitions and bulk chemical agents were found in the British Zone.[224]

Given that 119,910 tons of German CW munitions were found in the British Zone; 10,436 tons and 3,000 tons were transferred to the British from the American and French forces, respectively; 24,958 tons of excess U.S. Army stocks were "secretly" turned over to the British for disposal; and that 10,000 tons were shipped from the British Zone to the U.K. to augment the British stockpile (ostensibly for research purposes); it is concluded that approximately 148,304 tons of CW munitions were sea-disposed by the British between 1945 and 1947.

CW AMMUNITION SEA-DISPOSED BY THE U.K.
1945-47

Tons of CW Munitions found in the British Zone	119,910
Tons transferred from the American Zone	+ 10,436
Tons transferred from the French Zone	+ 3,000
Tons shipped to the U.K.	- 10,000
Excess U.S. stocks transferred for U.K. disposal	+24,958
TOTAL:	148,304

29 Jan 48   British Zone: CCG(BE) Zonal Demilitarisation Committee meeting #16

The 16th meeting of the Zonal Demilitarisation Committee of the Control Commission for Germany (British Element) was held at 1430 hours, Thursday, January 29, 1948 in the Tax House, Lübbecke. Brigadier J. N. D. Tyler, ACOS (Exec)B was in the Chair. No report was made regarding the disposition of CW.

1 Mar 48   16th of 24 UK Atlantic Sea Disposal Operations (1 ship: Harm Freitzen)

11 Mar 48   British Zone: CCG(BE) Zonal Demilitarisation Committee meeting #17

The 17th meeting of the Zonal Demilitarisation Committee of the Control Commission for Germany (British Element) was held on Thursday, March 11, 1948 in the Tax House, Lübbecke. No report was made regarding the disposition of CW.

22 Apr 48   British Zone: CCG(BE) Zonal Demilitarisation Committee meeting #18

The 18th meeting of the Zonal Demilitarisation Committee of the Control Commission for Germany (British Element) was held at 1430 hours, Thursday, April 22, 1948 in the Tax House, Lübbecke. Brigadier J. N. D. Tyler, ACOS (Exec) B was in the Chair. For the first time, Colonel J. H. S. Fea represented HQ, BAOR. Lt. Cdr. B. C. Ward, R.N. represented Naval Division. The combined Services Division (Army Branch) was not represented and no report was made regarding the disposition of CW. "The Committee agreed that there would be no objection to the discontinuation of the BAOR Monthly Disarmament Progress Report, if BAOR so desired."[225]

27 May 48   British Zone: CCG(BE) Zonal Demilitarisation Committee meeting #19

The 19th meeting of the Zonal Demilitarisation Committee of the Control Commission for Germany (British Element) was held at 1430 hours, Thursday, May 28, 1948 in the Tax House, Lübbecke. No report concerning the disposition of CW was made.

1 Jul 48   British Zone: CCG(BE) Zonal Demilitarisation Committee meeting #20

The 20th meeting of the Zonal Demilitarisation Committee of the Control Commission for Germany (British Element) was held at 1430 hours, Thursday, July 1, 1948 in the Tax House, Lübbecke. Brigadier J. N. D. Tyler, ACOS (Exec) B was in the Chair. No report was made concerning the disposition of CW.

24 Jul 48   American Zone: "Davey Jones Locker" Chemical Weapons Convoy #8 (DJL8)

The Philip Heiniken, a German freighter, was scuttled in the North Sea (at 62°57' N 01°32' E) containing 2,240 tons (2,000 long tons) of CW.

12 Aug 48   British Zone: CCG(BE) Zonal Demilitarisation Committee meeting #21

The 21st meeting of the Zonal Demilitarisation Committee of the Control Commission for Germany (British Element) was held at 1430 hours, Thursday, August 12, 1948 in the Tax House, Lübbecke. Lt.Col. B. S. Jerome, A/ACOS (Exec) B was in the Chair. That the Chairmanship of this Committee had been downgraded from a Brigadier to a Lieutenant Colonel is evidence of the Committee's diminished importance. As was the case since the Committee's sixth meeting on January 23, 1947, no report was made concerning the disposition of CW.

22 Aug 48   17th of 24 UK Atlantic Sea Disposal Operations (1 ship: Empire Success)

24 Aug 48   American Zone: "Davey Jones Locker" Chemical Weapons Convoy #9 (DJL9)

The Marcy, a German freighter, was scuttled in the North Sea (at 62°59' N 01°23' E) containing 2,800 tons (2,500 long tons) of CW. This was the last of eleven ships sunk in Operation Davey Jones Locker. 11 ships, with a combined total of 31,879.7 tons of chemical warfare ammunition, were scuttled by the United States in this 785-day operation. The U.S. dumping program exceeded the U.K. program by 175 days.

16 Sep 48   British Zone: CCG(BE) Zonal Demilitarisation Committee meeting #22

The 22nd meeting of the Zonal Demilitarisation Committee of the Control Commission for Germany (British Element) was held at 1430 hours, Thursday, September 16, 1948 in the Tax House, Lübbecke. No report on the disposition of CW was made.

22 Sep 48   18th of 24 UK Atlantic Sea Disposal Operations (1 ship: Miervaldis)

21 Oct 48   British Zone: CCG(BE) Zonal Demilitarisation Committee meeting #23

The 23rd meeting of the Zonal Demilitarisation Committee of the Control Commission for Germany (British Element) was held at 1430 hours, Thursday, October 21, 1948 in the Tax House, Lübbecke. No report was made concerning the disposition of CW.

25 Nov 48   British Zone: CCG(BE) Zonal Demilitarisation Committee meeting #24

The 24th meeting of the Zonal Demilitarisation Committee of the Control Commission for Germany (British Element) was held at 1430 hours, Thursday, November 25, 1948 in the Tax House, Lübbecke. No report was made concerning the disposition of CW.

13 Jan 49   British Zone: CCG(BE) Zonal
Demilitarisation Committee meeting #25

The 25th meeting of the Zonal Demilitarisation
Committee of the Control Commission for Germany
(British Element) was held at 1430 hours, Thursday,
January 13, 1949 in the Tax House, Lübbecke. No report
was made concerning the disposition of CW.

3 Mar 49   British Zone: CCG(BE) Zonal
Demilitarisation Committee meeting #26

The 26th meeting of the Zonal Demilitarisation
Committee of the Control Commission for Germany
(British Element) was held at 1430 hours, Thursday,
March 3, 1949 in the Tax House, Lübbecke. No report
was made concerning the disposition of CW.

7 Apr 49   British Zone: CCG(BE) Zonal
Demilitarisation Committee meeting #27

The 27th meeting of the Zonal Demilitarisation
Committee of the Control Commission for Germany
(British Element) was held at 1430 hours, Thursday,
April 7, 1949 in the Tax House, Lübbecke. No report was
made concerning the disposition of CW.

19 May 49   British Zone: CCG(BE) Zonal
Demilitarisation Committee meeting #28

The 28th meeting of the Zonal Demilitarisation
Committee of the Control Commission for Germany
(British Element) was held at 1430 hours, Thursday, May
19, 1949 in the Tax House, Lübbecke. No report was
made concerning the disposition of CW.

20 Jun 49   19th of 24 UK Atlantic Sea Disposal
             Operations (1 ship: Empire Connyngham)

14 Jul 49   British Zone: CCG(BE) Zonal
             Demilitarisation Committee meeting #29

The 29th meeting of the Zonal Demilitarisation
Committee of the Control Commission for Germany
(British Element) was held at 1430 hours, Thursday, July
14, 1949 in the Tax House, Lübbecke. No report was
made concerning the disposition of CW.

1 Sep 49   British Zone: CCG(BE) Zonal
             Demilitarisation Committee meeting #30

The 30th meeting of the Zonal Demilitarisation
Committee of the Control Commission for Germany
(British Element) was held at 1430 hours, Thursday,
September 1, 1949 in the Tax House, Lübbecke. No report
was made concerning the disposition of CW.

6 Oct 49   British Zone: CCG(BE) Zonal
             Demilitarisation Committee meeting #31

The penultimate meeting of the British Zonal
Demilitarisation Committee commenced at 1430 hours,
Thursday, October 6, 1949 in the Tax House, Lübbecke.
No report on CM was made.

8 Dec 49   British Zone: CCG(BE) Zonal
             Demilitarisation Committee meeting #32

The 32d and final meeting of the British Zonal
Demilitarisation Committee commenced at 1430 hours,
Thursday, December 8, 1949 in the Tax House, Lübbecke.
Brigadier J.N.D. Tyler, ACOS (Exec) B was in the Chair.
Lt.Col. The Honourable M. G. Edwardes represented HQ,

BAOR and Lt.Cdr. A. W .M. Matthew, Royal Navy, represented FOCBNG. No report was made concerning the disposition of CW in the British Zone. It was decided the Demilitarisation Committee would meet in the future as required and not at regular intervals.

27 Jul 55   20th of 24 UK Atlantic Sea Disposal Operations (1 ship: Empire Claire) (Operation Sandcastle)

30 May 56   21st of 24 UK Atlantic Sea Disposal Operations (1 ship: Vogtland) (Operation Sandcastle)

23 Jul 56   22nd of 24 UK Atlantic Sea Disposal Operations (1 ship: Kotka) (Operation Sandcastle)

Jun-Sep 56   23rd and 24th of 24 UK Atlantic Sea Disposal Operations (2 ships: Unknown).

Mar 60   German Disposal Operations

The two ships laden with 69,000 tabun shells which were scuttled in the Baltic's Little Belt before the end of the war, were surfaced. The 69,000 nerve gas shells were encapsulated in concrete and re-dumped in the Bay of Biscay.

1989-1990   Suspected Russian disposal operations in the Baltic.

APPENDIX D
ALPHABETICAL LIST OF THE
SCUTTLED
SHIPS

			Date		
		Tonnage	Scuttled	Latitude	Longitude
1	ALCO BANNER	3,097	14-Jul-46	58 18' 07" N	09 36' 05" E
2	ARTHUR SEWALL		12-Oct-46	Unk	Unk
3	BALKAN	3,500	17-Oct-45	58 16' N	09 26' E
4	BERLIN		31-May-46	57 08' 50" N	10 49' 12" E
5	BERNLEF / WAR OLIVE		14-Aug-45	56 10' 00" N	12 07' 00" E
6	BOTLEA / AFRICAN PRINCE		30-Dec-45	55 30.00 N	11 00.00 W
7	BRANDENBURG (VS-158)		Unk	Unk	Unk
8	BREMSE		Unk	57 52' 00" N	06 15' 00" E
9	CLAUS VON BEVERN (T-190)		16-Mar-46	57 52' 00" N	06 15' 00" E
10	DESSAU	6,000	17-May-47	Unk	Unk
11	DEUTSCHLAND	1,061	08-Sep-46	58 17' N	09 36' E
12	DORA OLDENDORFF		05-Feb-47	47 40.00 N	09 22.00 W
13	DRAU	8,000	17-Oct-45	58 16' N est.	09 26' E est.
14	DUBORG	3,500	04-Oct-45	58 14.5' N	09 31' E
15	EDITH HOWALDT	3,000	17-Oct-45	58 15' N	09 30' E
16	EIDER (Sperrbrecher 36)	5,000	12-Oct-46	58 17' N	09 38' E
17	EMMY FRIEDRICH	8,000	17-Oct-45	58 14' N	09 27' E
18	EMPIRE CLAIRE		27-Jul-55	56 30.00 N	12 00.00 W
19	EMPIRE CONNYNGHAM		20-Jun-49	47 52.00 N	08 51.00 W
20	EMPIRE CORMORANT		1-Oct-45	55 30.00 N	11 00.00 W
21	EMPIRE FAL		2-Jul-45	55 00.09 N	11 00.00 W
22	EMPIRE LARK		22-Jul-47	47 55.00 N	08 17.00 W
23	EMPIRE NUTFIELD		3-Sep-46	48 03.00 N	08 09.00 W
24	EMPIRE PEACOCK		25-Aug-46	47 57.00 N	08 33.24 W
25	EMPIRE SEVERN	4,700	12-Oct-46	58 18' 30" N	09 37' E
26	EMPIRE SIMBA		11-Sep-45	55 30.00 N	11 00.00 W
27	EMPIRE SUCCESS		22-Aug-48	47 16.30 N	09 24.00 W
28	EMPIRE WOODLARK		2-Nov-46	59 00.00 N	07 40.00 W
29	ERIKA SCHUNEMANN		Unk	Unk	Unk
30	F-192		Unk	58 08' 00" N	10 52' 00" E
31	FALKENFELS	10,000	16-Mar-46	58 14' N est.	09 24' E est.
32	FECHENHEIM	8,036	16-Mar-46	58 14' N est.	09 24' E est.
33	FREIBURG	6,500	13-Jul-46	Unk	Unk
34	GEMLOCK		Unk	Unk	Unk
35	GEORGE HAWLEY	1,120	30-Jun-47	58 18' 05" N	09 38' 00" E
36	GERTRUD FRITZEN	4,500	13-Jul-46	Unk	Unk
37	H.C. HORN		26-May-46	58 08' 30" N	10 50' 00" E
38	HARM FREITZEN		01-Mar-48	47 55.00 N	08 58.00 W
39	HELGOLAND		1948	Unk	Unk
40	HERBERT NORKUS		1947	Unk	Unk
41	HUGO OLDENDORF	8,037	16-Mar-46	58 14' N est.	09 24' E est.
42	JAMES HARROD	3,360	20-Jun-47	58 16' 00" N	09 33' 00" E
43	JAMES OTIS	4,091	30-Aug-46	58 16' 00" N	09 32' 00" E
44	JAMES SEWELL	4,480	06-Jun-47	58 15' 02" N	09 30' 06" E
45	JAMES W. NESMITH		03-Jun-46	Unk	Unk
46	JAN WELLENS		Oct-45	58 00' 00" N	09 30' 00" E
47	JANTJE FRITZEN	6,600	17-Nov-45	58 15' N	09 30' E
48	KARL LEONHARDT	8,000	16-Mar-46	58 14' N	09 24' E
49	KINDERSLEY		01-Oct-46	47 54.00 N	08 21.00 W
50	KOTKA		23-Jul-56	56 31.00 N	12 05.00 W

		Tonnage	Date Scuttled	Latitude	Longitude
51	KRYSSER LEIPZIG	1,000	16-Dec-46	57 52.011 N	06 15.747 E
52	KSB-1		Unk	Unk	Unk
53	KSB-13		Unk	Unk	Unk
54	LANARK		11-Nov-46	48 00.00 N	08 21.00 W
55	LEIGHTON		09-Aug-47	56 22.00 N	09 27.00 W
56	LOTTE		26-Mar-46	58 19' 00" N	09 40' 00" E
57	LOUISE SCHROEDER	1,800	04-Oct-45	58 15' N	09 27' E
58	LUDWIGSHAVEN	1,721	12-Oct-46	58 17' N	09 38' E
59	M-16		18-May-46	58 10' 12" N	10 42' 24" E
60	M-280		26-Jul-46	57 40' 00" N	06 30' 00" E
61	M-522		18-May-46	58 10' 12" N	10 40' 48" E
62	MARCY	2,800	24-Aug-48	62 59' 00" N	01 23' 00" E
63	MARGO		03-Nov-47	47 36.00 N	09 31.00 W
64	MIERVALDIS		22-Sep-48	47 23.00 N	09 24.00 W
65	MONTE PASCOAL	6,000	21-Dec-46	58 10.31 N	10 46.13 E
66	NESBITT	6,720	18-Jul-47	58 15' 00" N	09 30' 00" E
67	OCEAN TRANSPORT 2		Unk	Unk	Unk
68	ODERSTROM	2,465	17-Oct-45	58 16' N est.	09 26' E est.
69	OLGA SIEMERS	5,000	17-Oct-45	58 16' N est.	09 26' E est.
70	PATAGONIA	8,000	04-Oct-45	58 15' N	09 35' E
71	PHILIP HEINIKEN	2,240	24-Jul-48	62 57' 00" N	01 32' 00" E
72	PILLAU	1,800	04-Oct-45	58 15' N est.	09 30' E est.
73	RHON	1,061	08-Sep-46	58 17' N	09 36' 45" E
74	S-7		02-May-46	58 09' 00" N	10 50' 00" E
75	S-9		02-May-46	58 09' 00" N	10 51' 00" E
76	S-12		02-May-46	58 09' 00" N	10 52' 00" E
77	SCHWABENLAND	6,000	06-Jun-47	58 10.22 N	10 45.24 E
78	SESOSTRIS	2,000	17-Nov-45	58 18.315 N	09 41.057 E
79	SPERRBRECHER	1,511	01-Jul-46	58 14' 00" N	09 15' 00" E
80	T-21		16-Dec-46	57 52' 00" N	06 15' 00" E
81	T-37		26-Jul-46	57 40' 00" N	06 30' 00" E
82	T-38		10-May-46	58 07' 48" N	10 46' 30" E
83	T-39		10-May-46	58 08' 12" N	10 47' 48" E
84	T-63	6,000	21-Dec-46	Unk	Unk
85	T-65	1,709	01-Jul-46	58 17' 09" N	09 37' 01" E
86	T-156		03-May-45	Unk	Unk
87	TAGILA	2,600	17-Nov-45	58 15' N	09 30' E
88	TAURUS	1,000	17-Nov-45	58 16.009 N	09 31.152 E
89	TF-1		02-May-46	58 09' 15" N	10 50' 30" E
90	THEDA FRITZEN	2,466	17-Nov-45	58 18' N	09 55' E
91	THORPE BAY		08-Sep-47	47 47.30 N	08 21.00 W
92	TRITON	2,000	04-Oct-45	58 15' N est.	09 30' E est.
93	TRUDE SCHUNEMANN	1,500	17-Oct-45	58 16' N est.	09 26' E est.
94	U-J 305	752	02-Jul-46	58 16' 04" N	09 29' 00" E
95	VOGTLAND		30-May-56	56 30.00 N	12 00.00 W
96	WAIRUNA		30-Oct-45	55 30.00 N	11 00.00 W
97	Z-29		16-Dec-46	57 52' 00" N	06 15' 00" E
98	Z-34		26-Mar-46	58 19' N est.	09 40' E est.
99	Unknown Ship Name #1		1956	56 00.00 N	10 00.00 W
100	Unknown Ship Name #2		1956	56 00.00 N	10 00.00 W

		Primary Source(s)
1	ALCO BANNER	U.S. Army Munitions Command, Edgewood Arsenal, Maryland, "Inventory", 25 Aug 70
2	ARTHUR SEWALL	Norway 2002, http://www.mariners-l.co.uk/LibshipsA.html
3	BALKAN	U.K. MoD, "Report of Sea Dumping of CW in the Skagerrak Waters", 1993
4	BERLIN	MoD 1993, Norway 2002
5	BERNLEF / WAR OLIVE	Norway 2002, http://www.mariners-l.co.uk/WWIStandardShipsWarJ.html, 2,579 G.T.
6	BOTLEA / AFRICAN PRINCE	U.K. MoD, DSC-Env1, "Standard Reply to Enquiries Re Sea Dumping of Munitions"
7	BRANDENBURG (VS-158)	Norway 2002
8	BREMSE	Norway 2002
9	CLAUS VON BEVERN (T-190)	MoD 1993, Norway 2002
10	DESSAU	U.K. MoD, "Report of Sea Dumping of CW in the Skagerrak Waters", 1993
11	DEUTSCHLAND	U.K. MoD, "Report of Sea Dumping of CW in the Skagerrak Waters", 1993
12	DORA OLDENDORFF	U.K. MoD, DSC-Env1, "Standard Reply to Enquiries Re Sea Dumping of Munitions"
13	DRAU	U.K. MoD, "Report of Sea Dumping of CW in the Skagerrak Waters", 1993
14	DUBORG	U.K. MoD, "Report of Sea Dumping of CW in the Skagerrak Waters", 1993
15	EDITH HOWALDT	U.K. MoD, "Report of Sea Dumping of CW in the Skagerrak Waters", 1993
16	EIDER (Sperrbrecher 36)	U.K. MoD, "Report of Sea Dumping of CW in the Skagerrak Waters", 1993
17	EMMY FRIEDRICH	U.K. MoD, "Report of Sea Dumping of CW in the Skagerrak Waters", 1993; Norway 2002
18	EMPIRE CLAIRE	U.K. MoD, DSC-Env1, "Standard Reply to Enquiries Re Sea Dumping of Munitions"
19	EMPIRE CONNYNGHAM	U.K. MoD, DSC-Env1, "Standard Reply to Enquiries Re Sea Dumping of Munitions"
20	EMPIRE CORMORANT	U.K. MoD, DSC-Env1, "Standard Reply to Enquiries Re Sea Dumping of Munitions"
21	EMPIRE FAL	U.K. MoD, DSC-Env1, "Standard Reply to Enquiries Re Sea Dumping of Munitions"
22	EMPIRE LARK	U.K. MoD, DSC-Env1, "Standard Reply to Enquiries Re Sea Dumping of Munitions"
23	EMPIRE NUTFIELD	U.K. MoD, DSC-Env1, "Standard Reply to Enquiries Re Sea Dumping of Munitions"
24	EMPIRE PEACOCK	U.K. MoD, DSC-Env1, "Standard Reply to Enquiries Re Sea Dumping of Munitions"
25	EMPIRE SEVERN	U.K. MoD, "Report of Sea Dumping of CW in the Skagerrak Waters", 1993
26	EMPIRE SIMBA	U.K. MoD, DSC-Env1, "Standard Reply to Enquiries Re Sea Dumping of Munitions"
27	EMPIRE SUCCESS	U.K. MoD, DSC-Env1, "Standard Reply to Enquiries Re Sea Dumping of Munitions"
28	EMPIRE WOODLARK	U.K. MoD, DSC-Env1, "Standard Reply to Enquiries Re Sea Dumping of Munitions"
29	ERIKA SCHUNEMANN	Norway 2002
30	F-192	Norway 2002
31	FALKENFELS	U.K. MoD, "Report of Sea Dumping of CW in the Skagerrak Waters", 1993
32	FECHENHEIM	U.K. MoD, "Report of Sea Dumping of CW in the Skagerrak Waters", 1993
33	FREIBURG	U.K. MoD, "Report of Sea Dumping of CW in the Skagerrak Waters", 1993
34	GEMLOCK	Norway 2002
35	GEORGE HAWLEY	U.S. Army Munitions Command, Edgewood Arsenal, Maryland, "Inventory", 25 Aug 70
36	GERTRUD FRITZEN	U.K. MoD, "Report of Sea Dumping of CW in the Skagerrak Waters", 1993
37	H.C. HORN	Norway 2002
38	HARM FREITZEN	U.K. MoD, DSC-Env1, "Standard Reply to Enquiries Re Sea Dumping of Munitions"
39	HELGOLAND	Agderposten, 11 April 1984 and Fonnum, p. 4, Norway 2002
40	HERBERT NORKUS	Vrak i Skagerrak, Norway 2002, http://www.esys.org/bigship/norcus.html
41	HUGO OLDENDORF	U.K. MoD, "Report of Sea Dumping of CW in the Skagerrak Waters", 1993
42	JAMES HARROD	U.S. Army Munitions Command, Edgewood Arsenal, Maryland, "Inventory", 25 Aug 70
43	JAMES OTIS	U.S. Army Munitions Command, Edgewood Arsenal, Maryland, "Inventory", 25 Aug 70
44	JAMES SEWALL	U.S. Army Munitions Command, Edgewood Arsenal, Maryland, "Inventory", 25 Aug 70
45	JAMES W. NESMITH	Norway 2002, http://www.mariners-l.co.uk/LibShipsJ-Ji.html
46	JAN WELLENS	U.K. MoD, "Report of Sea Dumping of CW in the Skagerrak Waters", 1993
47	JANTJE FRITZEN	U.K. MoD, "Report of Sea Dumping of CW in the Skagerrak Waters", 1993
48	KARL LEONHARDT	U.K. MoD, "Report of Sea Dumping of CW in the Skagerrak Waters", 1993; Norway 2002
49	KINDERSLEY	U.K. MoD, DSC-Env1, "Standard Reply to Enquiries Re Sea Dumping of Munitions"
50	KOTKA	U.K. MoD, DSC-Env1, "Standard Reply to Enquiries Re Sea Dumping of Munitions"

		Primary Source(s)
51	KRYSSER LEIPZIG	MoD 1993, 1984 and Fonnum 1993, p. 14, Norway 2002: 57 52.011N, 06 15.747E, Kriegsmarine
52	KSB-1	Vrak i Skagerrak
53	KSB-13	Norway 2002
54	LANARK	U.K. MoD, DSC-Env1, "Standard Reply to Enquiries Re Sea Dumping of Munitions"
55	LEIGHTON	U.K. MoD, DSC-Env1, "Standard Reply to Enquiries Re Sea Dumping of Munitions"
56	LOTTE	Fonnum 1993, p. 12, Norway 2002
57	LOUISE SCHROEDER	U.K. MoD, "Report of Sea Dumping of CW in the Skagerrak Waters", 1993
58	LUDWIGSHAVEN	U.K. MoD, "Report of Sea Dumping of CW in the Skagerrak Waters", 1993
59	M-16	MoD 1993, Norway 2002
60	M-280	Norway 2002, http://www.german-navy.de/kriegsmarine/ships/minehunter/mboot40/ships.html
61	M-522	MoD 1993, Norway 2002, NO 2002 shows 58 10', 10 42'E
62	MARCY	U.S. Army Munitions Command, Edgewood Arsenal, Maryland, "Inventory", 25 Aug 70
63	MARGO	U.K. MoD, DSC-Env1, "Standard Reply to Enquiries Re Sea Dumping of Munitions"
64	MIERVALDIS	U.K. MoD, DSC-Env1, "Standard Reply to Enquiries Re Sea Dumping of Munitions"
65	MONTE PASCOAL	U.K. MoD, "Report of Sea Dumping of CW in the Skagerrak Waters", 1993; Norway 2002
66	NESBITT	U.S. Army Munitions Command, Edgewood Arsenal, Maryland, "Inventory", 25 Aug 70
67	OCEAN TRANSPORT 2	Norway 2002
68	ODERSTROM	U.K. MoD, "Report of Sea Dumping of CW in the Skagerrak Waters", 1993
69	OLGA SIEMERS	U.K. MoD, "Report of Sea Dumping of CW in the Skagerrak Waters", 1993
70	PATAGONIA	U.K. MoD, "Report of Sea Dumping of CW in the Skagerrak Waters", 1993
71	PHILIP HEINIKEN	U.S. Army Munitions Command, Edgewood Arsenal, Maryland, "Inventory", 25 Aug 70
72	PILLAU	U.K. MoD, "Report of Sea Dumping of CW in the Skagerrak Waters", 1993
73	RHON	U.K. MoD, "Report of Sea Dumping of CW in the Skagerrak Waters", 1993
74	S-7	Times, 4/5/92, MoD 1993 shows 58 08'30", 10 51'E, Norway 2002, Kriegsmarine verified
75	S-9	Times, 4/5/92, MoD 1993 shows 58 08'50", 10 52'30", Norway 2002, 18 Jan 46 per Kriegsmarine
76	S-12	Times, 4/5/92, MoD 1993 shows 58 09'10", 10 50'30"E, Norway 2002, 18 Jan 46 per Kriegsmarine
77	SCHWABENLAND	U.K. MoD, "Report of Sea Dumping of CW in the Skagerrak Waters", 1993; Norway 2002
78	SESOSTRIS	U.K. MoD, "Report of Sea Dumping of CW in the Skagerrak Waters", 1993; Norway 2002
79	SPERRBRECHER	U.S. Army Munitions Command, Edgewood Arsenal, Maryland, "Inventory", 25 Aug 70
80	T-21	Vrak i Skagerrak, Norway 2002, 10 Jun 46 Kriegsmarine
81	T-37	Fonnum 1993, p, 4; Agderposten, 11 Apr 84, Norway 2002, 26 Jul 46 per Kriegsmarine
82	T-38	Norway 2002, Kriegsmarine, MoD 1993
83	T-39	Norway 2002, Kriegsmarine, MoD 1993
84	T-63	U.K. MoD, "Report of Sea Dumping of CW in the Skagerrak Waters", 1993
85	T-65	U.S. Army Munitions Command, Edgewood Arsenal, Maryland, "Inventory", 25 Aug 70
86	T-156	Vrak i Skagerrak
87	TAGILA	U.K. MoD, "Report of Sea Dumping of CW in the Skagerrak Waters", 1993
88	TAURUS	U.K. MoD, "Report of Sea Dumping of CW in the Skagerrak Waters", 1993; Norway 2002
89	TF-1	Times, 4/5/92, Norway 2002, MoD 1993
90	THEDA FRITZEN	Norway 2002; Fonnum 1993
91	THORPE BAY	U.K. MoD, DSC-Env1, "Standard Reply to Enquiries Re Sea Dumping of Munitions"
92	TRITON	U.K. MoD, "Report of Sea Dumping of CW in the Skagerrak Waters", 1993
93	TRUDE SCHUNEMANN	U.K. MoD, "Report of Sea Dumping of CW in the Skagerrak Waters", 1993
94	U-J 305	U.S. Army Munitions Command, Edgewood Arsenal, Maryland, "Inventory", 25 Aug 70
95	VOGTLAND	U.K. MoD, DSC-Env1, "Standard Reply to Enquiries Re Sea Dumping of Munitions"
96	WAIRUNA	U.K. MoD, DSC-Env1, "Standard Reply to Enquiries Re Sea Dumping of Munitions"
97	Z-29	Fonnum 1993, p, 4; Agderposten, 11 Apr 84, Norway 2002
98	Z-34	MoD 1993, Norway 2002
99	Unknown Ship Name #1	U.K. MoD, DSC-Env1, "Standard Reply to Enquiries Re Sea Dumping of Munitions"
100	Unknown Ship Name #2	U.K. MoD, DSC-Env1, "Standard Reply to Enquiries Re Sea Dumping of Munitions"
	Fonnum 1993 =	Dr. Frode Fonnum, "An Investigation into the Sunken Chemical Ships", NDRE, Norway, 1993
	MoD 1993 =	U.K. MoD, "Report of Sea Dumping of CW in the Skagerrak Waters", 1993
	Norway 2002 =	NDRE FFI/Rapport-2002/04951, 10 December 2002, Appendix A: Location of Sunken Gas Ships

BIBLIOGRAPHY

Addendum to the History of Captured Enemy Toxic Munitions in the American Zone, European Command, June 1947 to August 1948 Inclusive (Concluding Phase). U.S. Army, Office of the Chief, Chemical Division (Colonel Charles E. Loucks, Chief, Chemical Division), Headquarters, European Command, January 1949.

Applied Science and Analysis (ASA), Inc., Newsletter 08-5, Recent Scientific and Political Developments Regarding Sea-Dumped Chemical Weapons in the Baltic Sea, John Hart (SIPRI) and Thomas Stock (Dynasafe). Presented at the International Seminar on Sea-Dumped Chemical Weapons: Perspectives of International Cooperation, Ministry of Foreign Affairs, Republic of Lithuania, 30 September-1 October 2008.

Assessing Potential Ocean Pollutants. U.S. National Academy of Sciences, Washington, D.C., 1975.

Babiyevskiy, Kirill Konstantinovich. "German Chemical Weapons Dumped in Baltic After World War II – Cooperation Urged in Assessing Danger." Moscow Rossiyskaya Gazeta. June 3, 1992.

Babiyevskiy, Kirill Konstantinovich. "Ticking Chemical Bombs." Peace Courier. Vol. 5, May 1992. (Helsinki, Finland)

Baltic Marine Environment Protection Commission – Helsinki Commission (HELCOM), Ad-Hoc Working Group on Dumped Chemical Munition (HELCOM CHEMU), Control Methods for the State of the Marine Environment in the Baltic Sea Concerning Dumped Chemical Munition. HELCOM CHEMU 1/5/1, Submitted by Russia. April 19, 1993.

Baltic Marine Environment Protection Commission – Helsinki Commission (HELCOM), Ad-Hoc Working Group on Dumped Chemical Munition (HELCOM CHEMU), Definition for Chemical Munition. HELCOM CHEMU 1/2, Submitted by Sweden. April 16, 1993.

Baltic Marine Environment Protection Commission – Helsinki Commission (HELCOM), Ad-Hoc Working Group on Dumped Chemical Munition (HELCOM CHEMU). Fate and Effects of Dumped Chemical Warfare Agents. HELCOM CHEMU 1/5, Submitted by Denmark. April 7, 1993.

Baltic Marine Environment Protection Commission – Helsinki Commission (HELCOM), Ad-Hoc Working Group on Dumped Chemical Munition (HELCOM CHEMU), Investigation on Restricted Areas of the Former USSR. HELCOM CHEMU 1/7/1, Submitted by Coalition Clean Baltic. April 20, 1993.

Baltic Marine Environment Protection Commission – Helsinki Commission (HELCOM), Ad-Hoc Working Group on Dumped Chemical Munition (HELCOM CHEMU), National Information on Dumped Chemical Munition. HELCOM CHEMU 1/3/1, Submitted by Latvia. April 5, 1993.

Baltic Marine Environment Protection Commission – Helsinki Commission (HELCOM), Ad-Hoc Working Group on Dumped Chemical Munition (HELCOM CHEMU), Preliminary Report on Dumped Chemical Munition. HELCOM CHEMU 1/3/2, Submitted by Denmark. April 7, 1993.

Baltic Marine Environment Protection Commission – Helsinki Commission (HELCOM), Ad-Hoc Working Group on Dumped Chemical Munition (HELCOM CHEMU), Report on Chemical Munitions Dumped in the Baltic Sea, Report to the 15th meeting of the Helsinki Commission, March 8-11, 1994, January 1994.

Baltic Marine Environment Protection Commission – Helsinki Commission (HELCOM), Ad-Hoc Working Group on Dumped Chemical Munition (HELCOM CHEMU), Report on the Availability of Correct Information on Dumped Chemical Munition on the Swedish Continental Shelf. HELCOM CHEMU 1/3, Submitted by Sweden. April 5, 1993.

Baltic Marine Environment Protection Commission – Helsinki Commission (HELCOM), Ad-Hoc Working Group on Dumped Chemical Munition (HELCOM CHEMU), The Marine Expedition to the Locations of Dumped Chemical Munitions - The Stages of the Work, Facilities, Equipment. HELCOM CHEMU 1/6/2, Submitted by Russia. April 19, 1993.

Baltic Marine Environment Protection Commission – Helsinki Commission (HELCOM), "The Baltic Marine Environment 1999-2002", 2003.

Bascom, Willard. "The Disposal of Waste in the Ocean." Scientific American, August 1974, 16-25.

Bethell, Nicholas. "London Paper Says Russia Pulled `Wool Over Americans Eyes' – Scientist Says Russians Still Defy Chemical Weapons Ban." The London Times. March 17, 1994.

Bishop, P.L. Marine Pollution and its Control. New York: McGraw-Hill Book Company, 1983.

Borg, Stefan. "Gaslarm på fiskebåt: Vi Har Blivit Blinda." Aftonbladet. March 23, 1984.

Borgese, Elisabeth Mann, and David Krieger. The Tides of Change: Peace, Pollution, and Potential of the Oceans. New York: Mason/Charter, 1975.

Brankowitz, William R., Chemical Weapons Movement History Compilation. U.S. Army, Aberdeen Proving Ground, Office of the Program Manager for Chemical Munitions (Demilitarization and Binary), Report No. SAPEO-CDE-87001, June 12, 1987.

Brankowitz, William R., Sea Dumping of Chemical Weapons. (Undated)

Brankowitz, William R. Summary of Some Chemical Munitions Sea Dumps by the United States. U.S. Army, Aberdeen Proving Ground, Office of the Program Manager for Chemical Munitions (Demilitarization and Binary), January 30, 1989.

Brubaker, Douglas. Marine Pollution and International Law: Principles and Practices. London: Belhaven Press, 1993.

Captured Enemy Material: Occupation Forces in Europe Series, 1945-46. U.S. Army, Office of the Chief Historian, European Command, Frankfurt-Am-Main, Germany, 1947.

Carter, L.J. "Nerve Gas Disposal." Science, Vol. 169, 1970.

Chemical and Biological Warfare. U.S. Congress, Hearing Before the Committee on Foreign Relations, U.S. Senate, April 30, 1969. (Washington, DC: U.S. Government Printing Office, 1969).

Chemical Munitions in the Southern and Western Baltic Sea – Compilation, Assessment, and Recommendations. German Federal Ministry of Transport, Federal Maritime and Hydrographic Agency, Federal/Länder Government Working Group. Hamburg: May 1993.

Chemical Reference Handbook. U.S. Army Field Manual (FM) 3-8, Washington, DC, January 31, 1974.

Chemische Kampfstoffmunition in der südlichen und westlichen Ostsee: Bestandsaufnahme, Bewertung und Empfehlungen. Federal Republic of Germany, Bundesamt für Seeschiffahrt und Hydrographie, Bericht der Bund/Länder-Arbeitsgruppe, Hamburg, 1993.

Chipman, N. Liability Risk and Insurance Considerations - Cost and Benefit Implications. Paper presented at the Comett II Seminar on "Risk Assessment of Military Waste Dumped on the Baltic Sea Floor", Kiel, Germany, June 1993.

Churchill, R.R., and A.V. Lowe. The Law of the Sea. Manchester University Press, London, 1983.

Clark, Robert Bernard. Marine Pollution (3d edition). Oxford: Oxford University Press, 1992.

Collins, W. Action Plan – Ocean Disposal Program. U.S. Army Chemical Research, Development, and Engineering Center (CRDEC), Aberdeen Proving Ground, Maryland. CRDEC White Paper, August 1989.

Compilation of U.S. Army Toxic Chemical Warfare Agent Production, Storage, Test, Disposal, and Decontamination Operations, 1940-1970. U.S. Army, AMCSA-N, November 1970 (declassified).

Convention on the High Seas (13 UST 2312, TIAS 5200, UNTS 82). By the President of the United States, 29 April 1958.

Convention on the Prohibition of Military or Any Other Hostile Use of Environmental Modification Techniques, Geneva, May 18, 1977.

Convention on the Prohibition of the Development, Production, Stockpiling, and Use of Chemical Weapons and Their Destruction, Paris, 1993.

Correll, Elisabeth. "Östersjöns Stridsgaser får Ligga." Dagens Nyheter. June 26, 1992.

Dahl, T. "Förbjudet att Dumpa Stridsgas." Stockholm Svenska Dagbladet. December 13, 1985.

Dagens Nyheter, "Inga Tecken på Läckande Nervgas (No Sign of Leaking Nerve Gas)", August 18, 1992.

Dagens Nyheter, "Senapsgas i Östersjön: fiskare svårt skadade", June 19, 1969.

Dagens Nyheter, "Trålare fick upp gasbomb", June 2, 1991.

Danish Ministry of the Environment, National Environmental Protection Agency, Copenhagen, Update of Report Dated 7 May 1985 Concerning Environmental, Health and Safety Aspects Connected with the Dumping of War Gas Ammunition in the Waters Around Denmark, December 1, 1992.

Danish National Environmental Protection Agency, Copenhagen, Detailed Report on the Environmental, Health, and Security Issues Associated with the Dumping of Poison Gas Ammunition in Danish Waters, May 7, 1985.

Declaration on the Human Environment (26 Principles). The Stockholm Conference on the Human Environment, June 5-16, 1972.

Delayed Toxic Effects of Chemical Warfare Agents. Stockholm International Peace Research Institute (SIPRI). Almqvist & Wiksell International, Stockholm and New York, June 1975.

Der Spiegel, "Zurück ins Meer (Back in the Sea)", No. 12, March 18, 1985.

Der Tagesspiegel, "Hennig: Giftgasgranaten aus der Ostsee Bergen", March 2, 1992.

De Standard, "Defensie en Leefmilieu Moeten Munitieveld in Noordzee Ruimen (Chemical Munitions in the North Sea Just Outside Zeebrugge)", December 1, 1989.

Detailed Report on the Environmental, Health, and Security Issues Associated with the Dumping of Poison Gas Ammunition in Danish Waters. Denmark, Ministry of the Environment, National Environmental Protection Agency, Copenhagen, May 7, 1985.

Disposal of Chemical Weapons – Alternative Technologies. U.S. Congress, Office of Technology Assessment, Background Paper No. OTA-BP-O-95. Washington, DC: U. S. Government Printing Office, June 1992.

Disposal of Poisonous Gases. U.S. Congress, Hearing Before the Subcommittee on Fisheries and Wildlife Conservation, Committee on Merchant Marine and Fisheries, U.S. House of Representatives, 91st Congress, May 1969. (Washington, DC: U.S. Government Printing Office, 1969).

Dixon, Trevor R., and T. J. Dixon. "Munitions in British Coastal Waters." Marine Pollution Bulletin, Vol. 10, December 1979.

Donnelly, Paul J. Environmental Fate of Chemical Agents in Deep Seawater: Literature Review. U.S. Army, Lieutenant Colonel Paul J. Donnelly, Chemical Corps, August 3, 1990.

Douglas, Mary and Aaron Wildavsky. Risk and Culture: An Essay on the Selection of Technical and Environmental Dangers, University of California Press, Berkeley and Los Angeles, 1982.

Dr. A. H. Heineken Foundation for the Environment. Dumped Chemical Weapons in the Sea – Options. Professor Dr. Egbert K. Duursma, ed. Amsterdam, Netherlands, June 1999.

Dumped Chemical Weapons in the Sea – Options. Professor Dr. Egbert K. Duursma, ed. Dr. A. H. Heineken Foundation for the Environment, Amsterdam, Netherlands, June 1999.

Dumping in the Irish Sea. Letter from J. M. Stuart, U.K. Ministry of Defence to Dr. D.F. Shaw, Chairman, Irish Sea Forum, University of Liverpool, June 29, 1995.

Dumping of War Gases After the Last War. Norway, Norwegian Department of the Environment, Oslo, July 12, 1986.

Duursma, Professor Dr. Egbert K. ed. Dumped Chemical Weapons in the Sea – Options. Dr. A. H. Heineken Foundation for the Environment, Amsterdam, Netherlands, June 1999.

Duve, Nils. "Senapsgasen fra Anden Verdenskrig Koster Dyrt: Fisk for ca. 2 Mio er Kasseret på Bornholm på Kun Fire Måneder." Havfiskeren. May 2, 1991.

Encyclopedia of Oceanography. Encyclopedia of Earth Sciences, Vol. I., R.W. Fairbridge, ed. New York: Reinhold Publishing Corporation. 1966.

Enemy Tactics in Chemical Warfare. Special Series No, 24, Military Intelligence Division, War Department. Washington, D.C., September 1, 1944.

Engzell, Bo. "Senapsgas på drift gör fisket riskfyllt." Dagens Nyheter. March 26, 1984.

Environmental Defense Fund Letter, ed. Norma H. Watson. Environmental Defense Fund (EDF). Vol. 24, No. 5, September 1993.

Estimate of Fate and Hazard of Airplane Bombs, Charged Mustard Gas, When Sunk at Sea. U.S. Army, Memorandum from Colonel John R. Wood, Chief, Medical Division to Colonel E. C. Wallington, Chief of Chemical Corps, Washington, DC, February 10, 1947.

Examination of Chemical Warfare Ammunition in the Norwegian Sea – Possible Environmental Impact. North Atlantic Treaty Organization (NATO), Office of the Assistant Secretary General for Scientific and Environmental Affairs, Committee on the Challenges of Modern Society (CCMS). Paper presented by Professor Frode Fonnum (Norway) at the third meeting of the CCMS Pilot Study on Cross-Border Environmental Problems Emanating from Defence-Related Installations and Activities held at the Federal Armed Forces Defence Science Agency in Munster, Germany, November 22-24, 1993.

Fædrelandsvennen, "48,000 Tonn Giftgass Skal Senkes I Skagerakk (48,000 Tons Poisonous Gas Scuttled in Skagerakk)", July 2, 1946.

Fedorov, Lev Aleksandrovich and Vil Mirzayanov. "Article Reveals Details of Chemical Weapons Production – We Waged Chemical Warfare on Our Own Territory." Moscow Nezavisimaya Gazeta. October 30, 1992.

Fedorov, Lev Aleksandrovich. "The Myths and Legends of Chemical Disarmament – Fedorov Comments Further on CW Allegations, Sees Ecological Problems, Military `Crimes'." Moscow Izvestiya. December 3, 1992.

Fedorov, Lev Aleksandrovich and Vladimir Voronov. "Details of Soviet Chemical Weapons Program From Declassified Archives – Chemical Weapons or Chemical War." Moscow Khimiya I Zhizn. July 1993.

Fedorov, Lev Aleksandrovich. "Fedorov Charges Official Irresponsibility – Victims of Lawlessness: National Security in the Chemical Interior, A Couple of Questions for the Military-Chemical Complex."

Fedorov, Lev Aleksandrovich. "Investigation for Rossiya: The Chemical Death Complex." Moscow Rossiya. December 8-14, 1993.

Fedorov, Lev Aleksandrovich. "Fedorov Urges CWC Ratification – After the Panic. Apropos Ratification of the Chemical Disarmament Convention." Moscow Segodnya. July 5, 1994.

Fedorov, Lev Aleksandrovich. "Chemical Weapons in Russia: History, Ecology, Politics." Moscow Khimicheskoye Oruzhiye V Rossii: Istoriya, Ekologiya, Politika, July 27, 1994.

Fedorov, Lev Aleksandrovich. "Expert Eyes Russian Chemical Arsenals – Weapons: We Were Preparing for an All-Out Chemical War." Moscow Obshchaya Gazeta. January 26, 1995.

Feshback, Murray and Alfred Friendly, Jr. Ecocide in the USSR – Health and Nature under Siege. New York: Basic Books, 1992.

Final Report: Captured Toxic Ammunition Program. U.S. Army, Headquarters, Continental Base Section, European Command, Chemical Section (Colonel Walter A. Guild, Chemical Officer), May 22, 1947.

Final Report of Bari Mustard Casualties. U.S. Army, Lieutenant Colonel Stewart Alexander, Medical Corps, June 20, 1944.

Flensburger Tageblatt, "Keine Gefahr Durch Chemische Kampfstoffe im Kleinen Belt?", October 24, 1984.

Flournoy, Michele A. and Lee A. Feinstein. Chemical Weapons Facts. Factsheet prepared by the Center for Defense Information, Washington, DC, June 1986.

Fokin, Aleksandr Vasilyevich and Kirill Konstantinovich Babiyevskiy. "CW Destruction: Techniques, Ecological Dangers." Moscow Priroda. May 1992.

Fonnum, Frode. An Investigation into the Sunken Chemical Ships Outside the Norwegian Coast. Norwegian Defence Research Establishment (NDRE)/Forsvarets Forskningsinstitutt (FFI), Kjeller, Norway.

Fonnum, Frode. Examination of Chemical Warfare (CW) Ammunition in the Norwegian Sea – Possible Environmental Impact. North Atlantic Treaty Organization (NATO), Office of the Assistant Secretary General for Scientific and Environmental Affairs, Committee on the Challenges of Modern Society (CCMS). Paper presented at the third meeting of the CCMS Pilot Study on Cross-Border Environmental Problems Emanating from Defence-Related Installations and Activities held at the Federal Armed Forces Defence Science Agency in Munster, Germany, November 22-24, 1993.

Forum Skagerrak. Vrak i Skagerrak. Sammanfattning av kunskaperna kring miljöriskerna med läckande vrak i Skagerrak. 2006.

Fourth Report on International Liability for Injurious Consequences Arising Out of Acts Not Prohibited by International Law. United Nations. Document A/CN.4/373.

Frankfurter Allgemeine Zeitung, "Aussenminister Streben Einen `Rat der Ostseestaaten' an Koordinierung der Regionalen Zusammenarbeit: Initiative Genschers und des Dänen Elleman-Jensen", March 5, 1992.

Frankfurter Allgemeine Zeitung, "Giftgasblase in der Ostsee Entdeckt", March 2, 1992.

Frankfurter Rundschau, "Entwarnung für Bornholm", March 16, 1992.

Frankfurter Rundschau, "Giftmüll in Ostsee Ungeheure Gefahr: Warnung Vor 1945 Versenkten Chemiewaffen", Vol. 48, No. 52, March 2, 1992.

Franson, Johan and Kaj Janerus. "Ang Vissa Fartyg Sänkta Med Last Av Stridsgas i Skagerrak." Sjöfartsverket. October 26, 1990.

Gamillscheg, Hannes. "Altes Giftgas Bedroht Skandinaviens Kusten (Poison Gas Threatens Scandinavia's Coasts)." Frankfurter Rundschau. October 23, 1990.

Gamillscheg, Hannes. "Im Blickpunkt: Atommüll und Giftgas: Skandinavier Leben in Angst." Frankfurter Rundschau. March 10, 1992.

Gardiner, William P. Ocean Dumping. Memorandum regarding ocean dumping issued by the Department of the Army, September 21, 1972.

German Chemical Warfare Materiel. U.S. Army, Headquarters European Theatre of Operations (ETO), Office of the Chief Chemical Warfare Officer, Intelligence Division, undated, declassified March 18, 1991.

German Federal Ministry of Transport, Federal Maritime and Hydrographic Agency, Federal/Länder Government Working Group, Hamburg, Chemical Munitions in the Southern and Western Baltic Sea – Compilation, Assessment, and Recommendations, May 1993.

Goldberg, Edward D. The Oceans as Waste Space. Scripps Institution of Oceanography, The University of Chicago, 1985.

Greene, L. Wilson. "Documents Relating to the Capture of a German Gas Dump." Armed Forces Chemical Journal.

Gwyther, Matthew. "Deadly Harvest." The London Daily Telegraph Weekend Magazine. November 11, 1989.

Harris, B.L. Chemicals in War. Vol. 5, New York: Wiley, 1979.

Harris, R. and J. Paxman. A Higher Form of Killing: The Secret Story of Gas and Germ Warfare. London, 1982.

Helgen, Lars. "De jobbar på en tickande miljöbomb." Göteborgs-Posten. July 12, 1991.

Hellberg, Hans. "Livsfarligt Hämta Upp Dumpad Stridsgas." Miljö Aktuellt. September 2, 1992.

Helsinki Commission (HELCOM) – Baltic Marine Environment Protection Commission. Ad-Hoc Working Group on Dumped Chemical Munition (HELCOM CHEMU), Control Methods for the State of the Marine Environment in the Baltic Sea Concerning Dumped Chemical Munition. HELCOM CHEMU 1/5/1, Submitted by Russia. April 19, 1993.

Helsinki Commission (HELCOM) – Baltic Marine Environment Protection Commission. Ad-Hoc Working Group on Dumped Chemical Munition (HELCOM CHEMU), Definition for Chemical Munition. HELCOM CHEMU 1/2, Submitted by Sweden. April 16, 1993.

Helsinki Commission (HELCOM) – Baltic Marine Environment Protection Commission. Ad-Hoc Working Group on Dumped Chemical Munition (HELCOM CHEMU), Fate and Effects of Dumped Chemical Warfare Agents. HELCOM CHEMU 1/5, Submitted by Denmark. April 7, 1993.

Helsinki Commission (HELCOM) – Baltic Marine Environment Protection Commission. Ad-Hoc Working Group on Dumped Chemical Munition (HELCOM CHEMU), Investigation on Restricted Areas of the Former USSR. HELCOM CHEMU 1/7/1, Submitted by Coalition Clean Baltic. April 20, 1993.

Helsinki Commission (HELCOM) – Baltic Marine Environment Protection Commission. Ad-Hoc Working Group on Dumped Chemical Munition (HELCOM CHEMU), National Information on Dumped Chemical Munition. HELCOM CHEMU 1/3/1, Submitted by Latvia. April 5, 1993.

Helsinki Commission (HELCOM) – Baltic Marine Environment Protection Commission. Ad-Hoc Working Group on Dumped Chemical Munition (HELCOM CHEMU), Preliminary Report on Dumped Chemical Munition (Dumpsites, Type and Quantity). HELCOM CHEMU 1/3/2, Submitted by Denmark. April 7, 1993.

Helsinki Commission (HELCOM) – Baltic Marine Environment Protection Commission, Ad-Hoc Working Group on Dumped Chemical Munition (HELCOM CHEMU), Report on Chemical Munitions Dumped in the Baltic Sea, Report to the 15th meeting of the Helsinki Commission, March 8-11, 1994, January 1994

Helsinki Commission (HELCOM) – Baltic Marine Environment Protection Commission. Ad-Hoc Working Group on Dumped Chemical Munition (HELCOM CHEMU), Report on the Availability of Correct Information on Dumped Chemical Munition on the Swedish Continental Shelf. HELCOM CHEMU 1/3, Submitted by Sweden. April 5, 1993.

Helsinki Commission (HELCOM) – Baltic Marine Environment Protection Commission. Ad-Hoc Working Group on Dumped Chemical Munition (HELCOM CHEMU), The Marine Expedition to the Locations of Dumped Chemical Munitions - The Stages of the Work, Facilities, Equipment. HELCOM CHEMU 1/6/2, Submitted by Russia. April 19, 1993.

Helsinki Commission (HELCOM) – Baltic Marine Environment Protection Commission, "The Baltic Marine Environment 1999-2002", 2003.

Hencke, David. "Chemical Weapons Dump Forces Rerouting of Gas Pipeline: Time Bomb under the Sea." London The Guardian. March 28, 1995.

Hood, Donald Wilbur (ed.). Impingement of Man on the Oceans. New York: John Wiley, 1971.

International Implications of Dumping Poisonous Gas and Waste into Oceans. U.S. Congress, Hearings Before the Subcommittee on International Organizations and Movements, Committee on Foreign Affairs, U.S. House of Representatives, May 8, 13, 14, and 15, 1969. (Washington, DC: U.S. Government Printing Office, 1969).

Inventory of Toxic Chemical and Biological (Toxin) Agents, Munitions, Dumps, and Historical Summary. U.S. Army, Edgewood Arsenal unnumbered report, August 25, 1970 (declassified).

Investigation of Ships Carrying Chemical Ammunition Sunk in Norwegian Waters After World War II. Norway, Conference on Disarmament, June 26, 1990.

Investigation of Shipwrecks Containing Chemical Ammunition Dumped in Norwegian Waters After World War II. Norway, Norwegian Defense Department Research Institute, Report No. 89/6007, Oslo, June 26, 1990.

Isherwood, Julian. "Nazi Chemical Bomb in Baltic Blinds Fishermen." The London Daily Telegraph. March 24, 1984.

Jansons, Aris. "Chemical Weapons Buried in Baltic Discussed – Deadly Weapons in Sea Next To Us." Riga (Latvia) Latvijas Juanatne. August 12, 1993.

Kiev Zelenyy Svit, "Chemical Weapons Dumped in Ukraine Waters – The Black Days of the Black Sea", July-August 1993.

Kiev Zelenyy Svit, "Chemical Weapons Program Impacts Environment", August 1994.

Kiev Zelenyy Svit, "Seas Serve as Naval Toxic Dumping Grounds – The Black Days of the Black Sea", 1994.

Kiev Zelenyy Svit, "Ukraine as Chemical Weapon Burial Ground – Generals Who Command...Bacteria", July-August 1993.

Kleber, Brooks E. and Dale Birdsell. United States Army in World War II, The Technical Series. The Chemical Warfare Service: Chemicals in Combat. Office of the Chief of Military History, United States Army, Washington, D.C., 1966.

Knightley, Phillip, "Dumps of Death." The London Times. April 5, 1992.

Krohn, W. Axel. "The Challenge of Dumped Chemical Ammunition in the Baltic Sea." Security Dialogue. Vol. 25, 1994.

Kudryashov, Sergey. "WWII Lewisite Dumps in `Dangerous' Condition." Moscow Izvestiya. July 1, 1994.

Laurin, Fredrik. "Dumpad stridsgas hotar Skagerak." Stockholm Svenska Dagbladet. October 16, 1990.

Laurin, Fredrik. "Enough Poison Gas in the Baltic to Kill Off Entire Population of Europe." Stockholm Tidingarnas Telegrambyrå. March 22, 1992.

Laurin, Fredrik. "Fishermen Risk Cancer and Blindness." Stockholm Tidingarnas Telegrambyrå. March 22, 1992. (See also Aasted, Annet. "Fishermen Exposed to Mustard Gas Clinical Experiences and Cancer Risk Evaluation.")

Laurin, Fredrik. "Massdumpad Giftgas i Östersjön: Gasbomberna från Andra Världskriget Tio Gånger Fler än man Tidigare Känt Till." Stockholm Svenska Dagbladet. March 23, 1992.

Lessons Learned from the Destruction of the Chemical Weapons of the Japanese Imperial Forces, ed. H. Kurata. Stockholm International Peace Research Institute (SIPRI). 1980.

Laurin, Fredrik. "Nations Worried About Poison Gas in Baltic Sea – Ten Times as Many Gas Bombs From World War II as Previously Known." Stockholm Svenska Dagbladet. March 23, 1992.

Laurin, Fredrik. "Scandinavia's Underwater Time Bomb." The Bulletin of the Atomic Scientists. March 1991.

Laurin, Fredrik. "Stridsgas hotar Skagerak." Stockholm Svenska Dagbladet. October 15, 1990.

Lok, Joris Janssen. "Cleaning Up the CW Legacy." Jane's Defence Weekly. Vol. 20, No. 19, November 6, 1993.

London Daily Telegraph. "Baltic Mustard Gas Burns 2", August 2, 1969.

Loucks, Colonel Charles E. and 1st Lieutenant R. Bruce Elliott. "Disposal of Captured Chemical Warfare Material in the U.S. Zone of Germany." Armed Forces Chemical Journal. Vol. 3, No. 4, January 1949.

Loucks, Colonel Charles E. "The Chemical Division, European Command." Armed Forces Chemical Journal. Vol 3, No. 7, January 1950.

Marine Pollution Bulletin, "Mustard Gas in the Baltic Sea", (15), 1969.

Marine Pollution Bulletin, "World War Two Poisons", 7, (10), 1976.

Memorandum for Technical Director of Edgewood Arsenal from Chief, Systems Analysis Office, Subject: Sea Dumps other than Operation CHASE. U.S. Army Edgewood Arsenal, Aberdeen Proving Ground, Maryland, May 26, 1969.

Michaelis, Dr. Anthony. "200,000 Tons of Gas Dumped in Sea by Great Britain." The London Daily Telegraph. August 12, 1970.

Military Chemistry and Chemical Compounds. U.S. Army Field Manual (FM) 3-9/U.S. Air Force Regulation (AFR) 355-7, Washington, DC, October 1975.

Mirzayanov, Vil. "Free to Develop Chemical Weapons." The Wall Street Journal. May 25, 1994.

Moscow Interfax, "Ministry: 'Tonnes' of Chemical Weapons 'Buried' at Sea", December 7, 1995.

Moscow Interfax, "Scientist Threatens to Reveal Secrets of Chemical Weapons", January 24, 1994.

Moscow ITAR-TASS, "Chemical Weapons Dumping After WWII Reported", February 11, 1994.

Moscow ITAR-TASS, "Dangers of Toxic Agents in Baltic Sea Noted", April 11, 1992.

Muller, Cornelia. "Ostsee-Boden Mit Arsen Verseucht." Kieler Nachrichten. June 3, 1992.

Narewski, M., Visual Inspection of War Gas Deposits by ROV. Paper presented at the Comett II 6689 Seminar on "Risk Assessment of Military Waste Dumped on the Baltic Sea Floor", Kiel, Germany, June 1993.

National Report of the Russian Federation, Moscow, Complex Analysis of the Hazard Related to the Captured German Chemical Weapon Dumped in the Baltic Sea, 11 October 1993.

New Scientist, "Irish Mustard", Vol. 115, No. 1568, July 9, 1987.

Non-Stockpile Chemical Materiel Program Interim Survey and Analysis Report. U.S. Army, Office of the Program Manager for Non-Stockpile Chemical Materiel, U.S. Army Chemical Materiel Destruction Agency, April 1993.

Non-Stockpile Chemical Materiel Program: Survey and Analysis Report. U.S. Army Chemical Materiel Destruction Agency, Office of the Program Manager for Non-Stockpile Chemical Materiel, November 1993.

North Atlantic Treaty Organization (NATO), Office of the Assistant Secretary General for Scientific and Environmental Affairs, Committee on the Challenges of Modern Society (CCMS). "Examination of Chemical Warfare Ammunition in the Norwegian Sea – Possible Environmental Impact". A paper presented by Professor Frode Fonnum (Norway) at the third meeting of the CCMS Pilot Study on Cross-Border Environmental Problems Emanating from Defence-Related Installations and Activities held at the Federal Armed Forces Defence Science Agency in Munster, Germany, November 22-24, 1993.

North Atlantic Treaty Organization (NATO), Office of the Assistant Secretary General for Scientific and Environmental Affairs, Committee on the Challenges of Modern Society (CCMS), Pilot Study on Cross-Border Environmental Problems Emanating from Defence-Related Installations and Activities, "Final Report, Volume 2: Chemical Contamination", August 1995.

North Atlantic Treaty Organization (NATO), Office of the Assistant Secretary General for Scientific and Environmental Affairs, Committee on the Challenges of Modern Society (CCMS), Pilot Study on Cross-Border Environmental Problems Emanating from Defence-Related Installations and Activities, "Summary of Proceedings and Decisions from the First Pilot Study Meeting in Oslo, Norway, February 11-12, 1993", February 12, 1993.

North Atlantic Treaty Organization (NATO), Office of the Assistant Secretary General for Scientific and Environmental Affairs, Committee on the Challenges of Modern Society (CCMS), Pilot Study on Cross-Border Environmental Problems Emanating from Defence-Related Installations and Activities, "Summary of Proceedings and Decisions from the Second Pilot Study Meeting in Kirkenes, June 1-2, 1993", June 2, 1993.

North Atlantic Treaty Organization (NATO), Office of the Assistant Secretary General for Scientific and Environmental Affairs, Committee on the Challenges of Modern Society (CCMS), Pilot Study on Cross-Border Environmental Problems Emanating from Defence-Related Installations and Activities, "Summary of Proceedings and Decisions from the Third Pilot Study Meeting in Munster, Germany, 22-24 November 1993", November 24, 1993.

North Atlantic Treaty Organization (NATO), Office of the Assistant Secretary General for Scientific and Environmental Affairs, Committee on the Challenges of Modern Society (CCMS), Pilot Study on Cross-Border Environmental Problems Emanating from Defence-Related Installations and Activities, "Summary of Proceedings and Decisions from the Fourth Pilot Study Meeting in Istanbul, Turkey, 18-19 April 1994", April 19, 1994.

Nottingham, Judith. "The Dangers of Chemical and Biological Warfare." Ecologist. Vol. 2, No, 2, 1972.

Nottingham, Judith and John Cookson. A Survey of Chemical and Biological Warfare. New York: Monthly Review Press, 1971.

Norwegian Conference on Disarmament, Investigation of Ships Carrying Chemical Ammunition Sunk in Norwegian Waters After World War II, June 26, 1990.

Norwegian Defence Research Establishment (NDRE), Forsvarets Forskningsinstitutt (FFI), Kjeller, Norway, An Investigation into the Sunken Chemical Ships Outside the Norwegian Coast, Professor Frode Fonnum.

Norwegian Defence Research Establishment (NDRE), Forsvarets Forskningsinstitutt (FFI), Kjeller, Norway, Examination of CW Ammunition in the Norwegian Sea – Possible Environmental Impact. Paper presented by Professor Frode Fonnum at the third meeting of the NATO CCMS Pilot Study on Cross-Border Environmental Problems Emanating from Defence-Related Installations and Activities held at the Federal Armed Forces Defence Science Agency in Munster, Germany, November 22-24, 1993.

Norwegian Defence Research Establishment (NDRE), Forsvarets Forskningsinstitutt (FFI), Kjeller, Norway, FFI/RAPPORT-2002/04951, "Investigation and Risk Assessment of Ships Loaded with Chemical Ammunition Scuttled in Skagerrak", 10 December 2002.

Norwegian Defence Research Establishment (NDRE), Forsvarets Forskningsinstitutt (FFI), Kjeller, Norway, FFI/RAPPORT-2009/02294, "Kartlegging av vrak med HUGIN HUS I dumpefelt for kjemisk ammunisjon i Skagerrak", Petter Långstad, 30 December 2009.

Norwegian Defence Research Establishment (NDRE), Forsvarets Forskningsinstitutt (FFI), Kjeller, Norway, Report Number FFI-NOTAT-92/6010, John Aasulf Tørnes, "Investigation of Ships Carrying Chemical Ammunition Sunk in Norwegian Waters After World War II", December 15, 1992.

Norwegian Pollution Control Authority, Oslo, Norway, Report Number TA-1907/2002, John Aasulf Tørnes, "Investigation and Risk Assessment of Ships Loaded with Chemical Ammunition Scuttled in Skagerrak", 18 November 2002.

Oberholz, Andreas. "Auf dem Pulverfass (On the Powder Keg)." Natur. April 4, 1989.

Oberholz, Andreas. "Eines Tages Könnten Guftgasklumpen bis an Bornholms Strände Treiben." Frankfurter Allgemeine Sonntagszeitung. March 1, 1992.

Oberholz, Andreas. "Toedliche Gefahr Aus der Tiefe." Kommunalpolitische Blaetter. Vol. 2, 1992.

Ocean Disposal of Unserviceable Chemical Munitions. U.S. Congress, Hearings Before the Subcommittee on Oceanography, Committee on Merchant Marine and Fisheries, U.S. House of Representatives, 91st Congress, Serial No. 91-31. (Washington, DC: U.S. Government Printing Office, 1970).

Ocean Dumping Authorization. U.S. Congress, Hearing Before the Subcommittee on the Environment and the Atmosphere, Committee on Science and Technology, U.S. House of Representatives, April 20, 1978. (Washington, DC: U.S. Government Printing Office, 1978).

Ocean Dumping Enforcement and the Current Status of Research Efforts. U.S. Congress, Hearing Before the Subcommittee on Coast Guard and Navigation, Committee on Merchant Marine and Fisheries, U.S. House of Representatives, October 1, 1992. (Washington, DC: U.S. Government Printing Office, 1993).

Olofson, Sune. "Stridsgas Skadade Fiskare." Stockholm Svenska Dagbladet. March 23, 1984.

OSPAR Commission, "Assessment of the Impact of Dumped Conventional and Chemical Munitions (Update 2009)", Publication Number: 365/2008, 2009.

OSPAR Commission, "Convention-wide Practices and Procedures in Relation to Marine Dumped Chemical Weapons and Munitions (2004 Update)", 2004.

OSPAR Commission, "Overview of Past Dumping at Sea of Chemical Weapons and Munitions in the OSPAR Maritime Area", 2002.

OSPAR Commission, "Overview of Past Dumping at Sea of Chemical Weapons and Munitions in the OSPAR Maritime Area – 2005 (Revised)", 2005.

Perera, Judith and Andy Thomas. "Fishing Boats Dodge Mustard Gas in the Baltic." New Scientist. Vol. 116, No. 1580, 1987.

Piskariov-Vassilie, A. The Feasibility of Geoelectric Methods for the Dumped Military Ammunition Location. Paper presented at the Comett II Seminar on "Risk Assessment of Military Waste Dumped on the Baltic Sea Floor", Kiel, Germany, June 1993.

Pleshakov, Leonid. "Further Revelations on WWII Chemical Dumping in Baltic – Investigation: The Bomb on the Bottom." Moscow Ogonek. February 8, 1992.

Polyukhov, Alexander. "What to Do With the `Drowned Stuff'." New Times. Vol. 37, September 12-13, 1992.

Pontecorvo, Giulio. The New Order of the Oceans: The Advent of a Managed Environment. New York: Columbia University Press, 1986.

Report of the Disposal Hazards of Certain Chemical Warfare Agents and Munitions. U.S. National Academy of Sciences, Ad-Hoc Advisory Committee of the National Academy of Sciences, Washington, DC. (14 pages) June 24, 1969.

Report of the First Meeting of the Ad-Hoc Working Group on Dumped Chemical Munition. Baltic Marine Environment Protection Commission – Helsinki Commission (HELCOM), (HELCOM CHEMU), St. Petersburg, Russia, 19-21 April 1993. HELCOM CHEMU 1/8. April 19-21, 1993.

Report on Chemical Munitions Dumped in the Baltic Sea. Baltic Marine Environment Protection Commission – Helsinki Commission (HELCOM), Ad-Hoc Working Group on Dumped Chemical Munition (HELCOM CHEMU), Report to the 15th meeting of the Helsinki Commission, March 8-11, 1994, January 1994.

Report on Sea Dumping of Chemical Weapons by the United Kingdom in the Skaggerrak Waters Post World War Two. U.K. Ministry of Defence, Great Scotland Yard, London, September 1993.

Reuters, "40,000 Tonnes of Chemical Weapons Dumped in Baltic Sea", October 1, 1993.

Riese, Birgitt. "Discarded Munitions Now Recognized as Environmental Hazard." Düsseldorf VDI Nachrichten. April 1, 1994.

Robinson, Julian Perry and Richard Dean Burns. Chemical/Biological Warfare: A Selected Bibliography. Center for the Study of Armament and Disarmament, California State University, Los Angeles, 1979.

Robinson, Julian Perry. The Effects of Weapons on Ecosystems. Published for the United Nations Environment Program and the United Nations Centre for Disarmament, Pergamon Press, New York, 1979.

Ruivo, M. (ed.). Marine Pollution and Sea Life. London: Fishing News Books, 1972.

Russian Federation, National Report of the Russian Federation, Moscow, Complex Analysis of the Hazard Related to the Captured German Chemical Weapon Dumped in the Baltic Sea, 11 October 1993.

Special Study on the Sea Disposal of Chemical Munitions. U.S. Arms Control and Disarmament Agency, U.S. Department of State. January 19, 1993.

Steinert, Harald. "Giftgas und Munition auf dem Grund der Ostsee." Tagesspiegel. July 17, 1991.

"Stridsgas Lämnas i Havet." Dagens Nyheter. August 8, 1992.

Stock, Thomas, ed. Old and Abandoned Chemical Weapons under the Chemical Weapons Convention. Stockholm International Peace Research Institute (SIPRI). December 1994.

Stock, Thomas and Karlheinz Lohs, eds. The Challenge of Old Chemical Munitions and Toxic Armament Wastes., Stockholm International Peace Research Institute (SPIRI). Oxford University Press, 1997.

Stock, Thomas, "Sea-Dumped Chemical Weapons Under the CWC – More Questions Than Answers?", Discussion Paper at the 52nd Pugwash CBW Workshop, Noordwijk, The Netherlands, 17-18 March 2007.

Stockholm International Peace Research Institute (SIPRI), Chemical Weapons: Destruction and Conversion. Taylor & Francis Ltd., London, 1980.

Stockholm International Peace Research Institute (SIPRI), Delayed Toxic Effects of Chemical Warfare Agents. Almqvist & Wiksell International, Stockholm and New York, June 1975.

Stockholm International Peace Research Institute (SIPRI), SIPRI Yearbook 1994, Oxford University Press, 1994.

Stockholm International Peace Research Institute (SPIRI), The Challenge of Old Chemical Munitions and Toxic Armament Wastes, Thomas Stock and Karlheinz Lohs, Oxford University Press, 1997.

Stockholm International Peace Research Institute (SIPRI), The Problem of Chemical and Biological Warfare, Volume I, The Rise of CB Weapons, Almquist & Wiksell, Stockholm and Humanities Press, New York, 1971.

Stockholm International Peace Research Institute (SIPRI), The Problem of Chemical and Biological Warfare, Volume II, CB Weapons Today, Almquist & Wiksell, Stockholm and Humanities Press, New York, 1973.

Stockholm International Peace Research Institute (SIPRI), Thomas Stock, ed. Old and Abandoned Chemical Weapons under the Chemical Weapons Convention, December 1994.

Stockholm International Peace Research Institute (SIPRI), Weapons of Mass Destruction and the Environment, Taylor & Francis, Ltd., London, 1977.

Storage, Shipment, Handling, and Disposal of Chemical Agents and Hazardous Chemicals. U.S. Army Technical Manual (TM) 3-250, Washington, DC, March 7, 1969.

Süddeutsche Zeitung, "Keine Giftgasblase Südlich von Bornholm", March 16, 1992.

Süddeutsche Zeitung, "Krause Lässt Giftgas in der Ostsee Untersuchen", March 11, 1992.

Süddeutsche Zeitung, "Luftblase in der Ostsee Enthält Vermutlich Giftgas", March 2, 1992.

Süddeutsche Zeitung, "Sorge Urn Die Schmutzige Ostsee: Anlieger: Sauberes Meer Gemeinsame Aufgabe: Aussenminister Wollen Marktwirtschaftliche Wachstumszone für Alle Länder", March 7-8, 1992.

Summary of Proceedings and Decisions from the First Pilot Study Meeting in Oslo, Norway, February 11-12, 1993. NATO (North Atlantic Treaty Organization), Office of the Assistant Secretary General for Scientific and Environmental Affairs, Committee on the Challenges of Modern Society (CCMS), Pilot Study on Cross-Border Environmental Problems Emanating from Defence-Related Installations and Activities, February 12, 1993.

Summary of Proceedings and Decisions from the Second Pilot Study Meeting in Kirkenes, June 1-2, 1993. NATO (North Atlantic Treaty Organization), Office of the Assistant Secretary General for Scientific and Environmental Affairs, Committee on the Challenges of Modern Society (CCMS), Pilot Study on Cross-Border Environmental Problems Emanating from Defence-Related Installations and Activities, June 2, 1993.

Summary of Proceedings and Decisions from the Third Pilot Study Meeting in Munster, Germany, 22-24 November 1993. NATO (North Atlantic Treaty Organization), Office of the Assistant Secretary General for Scientific and Environmental Affairs, Committee on the Challenges of Modern Society (CCMS), Pilot Study on Cross-Border Environmental Problems Emanating from Defence-Related Installations and Activities, November 24, 1993.

Sundaram, Navina. "Versunkene Granaten. Ostsee: 5,000 Quadratkilometer Sind Giftverseucht." Die Zeit. April 13, 1984.

Svenska Dagbladet, "Låt Giftgasbomberna Ligga", March 24, 1992.

Svenska Dagbladet, "Mycket senapsgas i Östersjön men vattnet bryter nre giftet", July 30, 1969.

Svenska Dagbladet, "Nya hudskador i Oxelösund: senapsgas misstänkt orsak", September 10, 1969.

Svenska Dagbladet, "Ryssar Söker Dumpad Stridsgas (Russians Seek Dumped War Gas)", August 19, 1992.

Svenska Dagbladet, "Senapsgas skadade ytterligare två svårt", August 2, 1969.

Tereshkin, Viktor. "Fedorov on Abuses in Secrecy on CW – Chas Pik Investigation: Secrets of Chemical Warfare." St. Petersburg (Russia) Chas Pik. March 17, 1993.

Terry, Anthony and David Divine. "After 24 Years – Dumped War Gas Hits Holiday Beaches." The London Sunday Times. August 10, 1969.

The Baltimore Sun, "Aberdeen's Alternatives", December 5, 1987.

The Baltimore Sun, "Russia Still Doing Secret Work on Chemical Arms – Research Goes On As Government Seeks U.N. Ban", October 18, 1992.

The Detoxification and Natural Degradation of Chemical Warfare Agents, ed. Ralf Trapp, Stockholm International Peace Research Institute (SIPRI), 1985.

The History of Captured Enemy Toxic Munitions in the American Zone, European Theater, May 1945 to June 1947. U.S. Army European Command, Office of the Chief of Chemical Corps (Colonel H. M. Woodward, Chief of Chemical Corps).

The Journal of Industrial and Engineering Chemistry, "Statement of U.S. Army Colonel V. Abbott,

Acting Chief of the U.S. Army Corps of Engineers", Vol. 11, No. 9, September 1919.

The Law of the Sea, Article 210, Pollution by Dumping, United Nations Convention on the Law of the Sea, New York, 1983.

The London Sunday Times, "Red Dust – The Disaster Menacing Europe", July 7, 1991.

The London Times, "Chemical Warfare Ammunition", July 24, 1947.

The London Times, "Dangerous Cargo", March 10, 1960.

The London Times, "Disposal of Chemical Ammunition", September 17, 1948.

The London Times, "Gas Shells Held for 10 Years after War", August 14, 1969.

The London Times, "Gas Ship is Scuttled in Atlantic", August 19, 1970.

The London Times, "Mustard Gas is Dumped in Sea", April 17, 1965.

The London Times, "Poison-Gas Dumped in the Sea", September 7, 1945.

The Problem of Chemical and Biological Warfare, Volume I, The Rise of CB Weapons. Stockholm International Peace Research Institute (SIPRI), Almquist & Wiksell, Stockholm and Humanities Press, New York, 1971.

The Problem of Chemical and Biological Warfare, Volume II, CB Weapons Today. SIPRI (Stockholm International Peace Research Institute). Almquist & Wiksell, Stockholm and Humanities Press, New York, 1973.

The Scadinavian Times, "Lethal Leftovers", November 1970.

The State of the Marine Environment. United Nations Environment Program (UNEP) Regional Seas Reports and Studies, United Nations Joint Group of Experts on the Scientific Aspects of Marine Pollution (GESAMP), No. 39, 1990.

Tiwe, Eva. "Så Dumpades de Livsfarliga Gasen Vid Skånes Kust." Arbetet. March 29, 1989.

Tjernberg, Urban. "Inga Läckor Från Vrak Med Gas (No Leaks from Wreck Containing Gas)." Stockholm Svenska Dagbladet. October 10, 1992.

Tørnes, John Aasulf. Investigation of Ships Carrying Chemical Ammunition Sunk in Norwegian Waters After World War II. Norwegian Defence Research Establishment (NDRE)/Forsvarets Forskningsinstitutt (FFI), Kjeller, Norway. Report Number FFI-NOTAT-92/6010, December 15, 1992.

Treaty on the Prohibition of the Emplacement of Nuclear Weapons and Other Weapons of Mass Destruction on the Seabed and the Ocean Floor and the Subsoil Thereof, Signed in Washington, London, and Moscow, February 11, 1971.

U.K. Ministry of Defence (MoD), "Report on Sea Dumping of Chemical Weapons by the United Kingdom in the Skagerrak Waters Post World War Two", September 1993.

U.K. Ministry of Defence (MoD), "Standard Reply to Enquiries Re Sea Dumping of Munitions", R. Bowles, MoD-DSC-Env1, undated.

United Nations, "Fourth Report on International Liability for Injurious Consequences Arising Out of Acts Not Prohibited by International Law", Document A/CN.4/373.

United Nations, "The State of the Marine Environment", United Nations Environment Program (UNEP) Regional Seas Reports and Studies, United Nations Joint Group of Experts on the Scientific Aspects of Marine Pollution (GESAMP), No. 39, 1990.

Update of Report Dated 7 May 1985 Concerning Environmental, Health and Safety Aspects Connected with the Dumping of War Gas Ammunition in the Waters Around Denmark. Denmark, Ministry of the Environment, National Environmental Protection Agency, Copenhagen, December 1, 1992.

U.S. Arms Control and Disarmament Agency (ACDA), U.S. Department of State, Special Study on the Sea Disposal of Chemical Munitions, January 19, 1993.

U.S. Army, "Addendum to the History of Captured Enemy Toxic Munitions in the American Zone, European Command, June 1947 to August 1948", Office of the Chief, Chemical Division (Colonel Charles E. Loucks, Chief, Chemical Division), Headquarters, European Command, January 1949.

U.S. Army, "Captured Enemy Material", Occupational Forces in Europe Series, 1945-46, Office of the Chief Historian, European Command, 1947.

U.S. Army, "Chemical Weapons Movement History Compilation", Aberdeen Proving Ground, Office of the Program Manager for Chemical Munitions (Demilitarization and Binary), Report No. SAPEO-CDE-87001, June 12, 1987.

U.S. Army, "Compilation of U.S. Army Toxic Chemical Warfare Agent Production, Storage, Test, Disposal, and Decontamination Operations, 1940-1970", November 1970.

U.S. Army, "Inventory of Toxic Chemical and Biological (Toxin) Agents, Munitions, Dumps, and Historical Summary", Edgewood Arsenal, Maryland, 25 August 1970.

U.S. Army, "Non-Stockpile Chemical Materiel Program Interim Survey and Analysis Report", Office of the Program Manager for Non-Stockpile Chemical Materiel, U.S. Army Chemical Materiel Destruction Agency, April 1993.

U.S. Army, "Non-Stockpile Chemical Materiel Program: Survey and Analysis Report", U. S. Army Chemical Materiel Destruction Agency, Office of the Program Manager for Non-Stockpile Chemical Materiel, November 1993.

U.S. Army, "The History of Captured Enemy Toxic Munitions in the American Zone, European Theater, May 1945 to June 1947", Office of the Chief of Chemical Corps, European Command, 1947.

U.S. Congress, Congressional Research Service (CRS), "CRS Report for Congress: U.S. Disposal of Chemical Weapons in the Ocean: Background and Issues for Congress", January 3, 2007.

U.S. National Academy of Sciences, Ad-Hoc Advisory Committee of the National Academy of Sciences, "Report of the Disposal Hazards of Certain Chemical Warfare Agents and Munitions", Washington, DC, June 24, 1969.

U.S. National Academy of Sciences, "Assessing Potential Ocean Pollutants", Washington, D.C., 1975.

U.S. National Academy of Sciences, "Report of the Disposal Hazards of Certain Chemical Warfare Agents and Munitions", Ad-Hoc Advisory Committee of the National Academy of Sciences, Washington, DC, June 24, 1969.

U.S./U.S.S.R. Executive Bilateral Agreement Relating to Fisheries in the Northeastern Part of the Pacific Ocean Off the United States Coast, February 13, 1967.

Weapons of Mass Destruction and the Environment. SIPRI (Stockholm International Peace Research Institute). Taylor & Francis, Ltd., London, 1977.

Wuschluhn, Franz. "Die Wahrheit Über das Gift in der Ostsee." Hamburger Abendblatt. March 27, 1984.

Ziemke, Earl Frederick. Army Historical Series: The U.S. Army in the Occupation of Germany 1944-1946. Center of Military History, United States Army, Washington, D.C., 1975.

END NOTES

[1] Dumped Chemical Weapons in the Sea – Options, p. viii.

[2] Chemical Weapons Convention (CWC), Article II, Definitions and Criteria.

[3] Ibid.

[4] Ibid.

[5] Stockholm International Peace Research Institute (SIPRI), Old and Abandoned Chemical Weapons under the Chemical Weapons Convention, p. 1.

[6] Overview of Past Dumping at Sea of Chemical Weapons and Munitions in the OSPAR Maritime Area, p. 5.

[7] Ibid.

[8] Stockholm International Peace Research Institute, Delayed Toxic Effects of Chemical Warfare Agents, p. 4.

[9] Stockholm International Peace Research Institute, The Challenge of Old Chemical Munitions and Toxic Armament Wastes, p. 4.

[10] Stockholm International Peace Research Institute, The Rise of CB Weapons (Vol. 1), The Problem of Chemical and Biological Warfare, pp. 153-155 and Dumped Chemical Weapons in the Sea – Options, p. 3.

[11] Stockholm International Peace Research Institute, The Challenge of Old Chemical Munitions and Toxic Armament Wastes, p. 263.

[12] Ibid, p. 4.

[13] Ibid, p. 266.

[14] National Report of the Russian Federation, Complex Analysis of the Hazard Related to the Captured German Chemical Weapon (sic) Dumped in the Baltic Sea, pp. 4, 6. (12,035 tons of agents / 70,500 tons gross weight)

[15] Report to the 15th Meeting of the Helsinki Commission from the Ad-Hoc Working Group on Dumped Chemical Munition (HELCOM CHEMU), January 1994, p. 11.

[16] Stockholm International Peace Research Institute, The Challenge of Old Chemical Munitions and Toxic Armament Wastes 1997, p. 269.

[17] Ibid, p. 4.

[18] Knightley, Phillip, The London Times, April 5, 1992.

[19] Laurin, Frederik, Bulletin of the Atomic Scientists, March 1991.

[20] The U.S. Arms Control and Disarmament Agency (ACDA) was established on September 26, 1961 as an agency of the U.S. Department of State. On April 1, 1999, ACDA became the U.S. Department of State's Bureau of Arms Control.

[21] U.S. Army, U.S. Army Chemical Materiel Destruction Agency, Program Manager for Non-Stockpile Chemical Materiel, Interim Survey and Analysis Report on the Non-Stockpile Chemical Materiel Program, April 1993, p. 16.

[22] Stockholm International Peace Research Institute, The Challenge of Old Chemical Munitions and Toxic Armament Wastes, p. 266.

[23] Control Commission for Germany (British Element).

---

[24] An oblast is a political subdivision of a republic in the FSU.

[25] Stockholm International Peace Research Institute, The Challenge of Old Chemical Munitions and Toxic Armament Wastes, p. 4.

[26] A copy of the formerly TOP SECRET document is archived at the Franklin D. Roosevelt Presidential Library and Museum. (http://docs.fdrlibrary.marist.edu/PSF/BOX32/T298E01.H TML

[27] The boundaries of the four occupation zones established at Yalta generally followed the borders of the former German federal states. Only Prussia constituted an exception: it was dissolved altogether, and its territory was absorbed by the remaining German Länder in northern and north-western Germany. Prussia's former capital, Berlin, differed from the rest of Germany in that it was occupied by all four Allies – and thus had so-called Four Power status. The occupation zone of the United States consisted of the Land of Hesse, the northern half of the present-day Land of Baden-Württemberg, Bavaria, and the southern part of Greater Berlin. The British zone consisted of Schleswig-Holstein, Lower Saxony, North Rhine-Westphalia, and the western sector of Greater Berlin. The French were apportioned Rhineland-Palatinate, the Saarland--which later received a special status--the southern half of Baden-Württemberg, and the northern sector of Greater Berlin. The Soviet Union controlled Mecklenburg, Brandenburg, Saxony, Saxony-Anhalt, Thuringia, and the eastern sector of Greater Berlin, which constituted almost half the total area of the city.

[28] U.K. Ministry of Defence, "Report on Sea Dumping of Chemical Weapons by the United Kingdom in the Skagerrak Waters Post World War Two", 1993, p. 1.

[29] Appendix A to Section II, DSWV Liaison Letter No. 35, The Disposal of CW Ammunition and of Bulk Gases in Germany, undated.

[30] U.K. Ministry of Defence, "Report on Sea Dumping of Chemical Weapons by the United Kingdom in the Skagerrak Waters Post World War Two", 1993, p. 5.

[31] William Porter (1886-1973) was promoted to Major General and assigned as Chief, Chemical Warfare Service (CWS) on 2 June 1941. MG Porter retired from the Army in November 1945. Brigadier General Alden H. Waitt, who was serving at the time as Assistant Chief of the CWS for Field Operations, was promoted to Major General in November 1945 and became Chief, CWS.

[32] At the start of WWII, Colonel Waitt took over duty as Chief, Plans and Training Division, CWS. Promoted to Brigadier General, he then served as Executive Officer of the CWS, second only to MG Porter. MG Waitt became Chief, Chemical Corps in November 1945 and resigned in 1949 when implicated in an industry scandal. He was replaced by MG Tony McAuliffe, Infantry.

[33] Minutes of the 1st Meeting of the Continental Ammunition Dumping Committee held at HQ, 21st Army Group, July 15, 1945.

[34] U.K. Ministry of Defence, "Report on Sea Dumping of Chemical Weapons by the United Kingdom in the Skagerrak Waters Post World War Two", 1993, p. 4.

[35] Ibid.

[36] The seaport of Stavanger is on Stavanger Fjord in SW Norway, 190 miles WSW of Oslo. It had been seized by the Germans on April 9, 1940.

---

[37] The Kattegat is a broad arm of the North Sea, 40 to 70 miles wide, between Sweden on the east and Jutland, Denmark on the west, connecting with the North Sea through the Skagerrak and with the Baltic Sea through Öresund, the Great Belt, and the Little Belt.

[38] U.K. Ministry of Defence, "Report on Sea Dumping of Chemical Weapons by the United Kingdom in the Skagerrak Waters Post World War Two", 1993, p. 5.

[39] U.K. Ministry of Defence, Documents on British Policy Overseas, Series I, Volume I, No. 603.

[40] Norwegian Defence Research Establishment (NDRE), Forsvarets Forskningsinstitutt (FFI), An Investigation into the Sunken Chemical Ships Outside the Norwegian Coast, Professor Frode Fonnum, 1993, p. 1.

[41] U.K. Ministry of Defence, "Report on Sea Dumping of Chemical Weapons by the United Kingdom in the Skagerrak Waters Post World War Two", 1993, p. 4.

[42] Minutes of the 3rd Meeting of the Continental Ammunition Dumping Committee held at HQ, 21st Army Group, August 16, 1945.

[43] U.K. Ministry of Defence, "Report on Sea Dumping of Chemical Weapons by the United Kingdom in the Skagerrak Waters Post World War Two", 1993, p. 3.

[44] U.S. Memoranda Concerning Captured WWII CW Material, Appendix I to Special Study on the Sea Disposal of Chemical Munitions, United States Arms Control and Disarmament Agency (ACDA), January 19, 1993.

[45] Public Record Office, Foreign Office (FO) File 1005/1959, "15th Progress Report of the Control Commission for Germany (British Element) Covering the Period 1-30 September, 1945",

Appendix A.

[46] Ibid, Appendix B, para. 2a.

[47] U.S. Army, Captured Enemy Ammunition, Office of the Chief Historian, European Command, 1947, p. 28.

[48] Ibid.

[49] U.S. Army, Final Report, Captured Toxic Ammunition Program, HQ, Continental Base Section, European Command, Chemical Section, May 22, 1947, p. 1.

[50] The Skaggerak Strait is located between the North Sea and the Baltic Sea.

[51] Stockholm International Peace Research Institute (SIPRI), The Challenge of Old Chemical Munitions and Toxic Armament Wastes, p. 265.

[52] U.S. Arms Control and Disarmament Agency (ACDA), U.S. Department of State, Special Study on the Sea Disposal of Chemical Munitions, 1993, p. ii.

[53] National Report of the Russian Federation, Complex Analysis of the Hazard Related to the Captured German Chemical Weapons Dumped in the Baltic Sea, p. 5.

[54] Public Record Office, War Office (WO) File 32, "SECRET Letter from HQ, BAOR to The War Office, Subject: CW Ammunition Held by the French," February 27, 1946.

[55] Final Report, Captured Toxic Ammunition Program, HQ, Continental Base Section, European Command, Chemical Section, May 22, 1947, p. 1. This statement, among other references, corroborates the fact that U.S. sea disposal operations occurred before the first official "Operation Davey Jones Locker" convoy.

[56] The first attempt at creating a joint command in Europe was made on 15 March 1947 when the European Command (EUCOM) replaced USFET. The purpose of the reorganization was to place in the hands of a single commander responsibility for the conduct of military operations of the land, naval and air forces. Although EUCOM was planned as a joint command, it never truly became one. The EUCOM "component commands" as of 15 November 1947 were the: U.S. Army Europe (USAREUR); the U.S. Air Forces in Europe (USAFE); and the U.S. Naval Forces, Europe.

[57] The term "Third Reich" is used as a synonym for Nazi Germany and was introduced by Nazi

propaganda, which counted the Holy Roman Empire as the first Reich, the 1871 German Empire as the second, and its own regime as the third. This was done in order to suggest a return to alleged former German glory after the perceived failure of the 1919 Weimar Republic.

[58] In addition to thiodiglycol, the Orgazid plant at Ammendorf near Leipzig manufactured mustard and nitrogen mustard starting in 1939.

[59] The Ergethan plant at Stassfurth/Loderburg manufactured Arsinol, an arsenical mixture.

[60] The Raubkammer plant at Munster. Starting in 1938, efforts were made here to standardize the design of CN and H munitions. Intensive work as also carried out on DA, DC, and DM and experiments were run with GA. Filling trials with GA and viscous H were conducted in 1942.

[61] Stockholm International Peace Research Institute, The Challenge of Old Chemical Munitions and Toxic Armament Wastes, Chapter 6.

[62] According to the U.S. Army source document, there is evidence of additional 'poison gas' production at the following locations, although the type or identity of the gas was not

made clear: Allen, Birkirchen, Bohneland near Brandenburg, Deggendorf, Draisdorf, Dresden, Falkensee Island, Goostacht, Griesheim, Hallehalle, Hanau, Hasmerheim, Henndorf, Heybrech near Breslau, Horpolding, Kendorf, Ludwigferdin, Moosberg, Ratibor, Rauxel, Opladen, Siemenstadt, Stettin, Villionbach, and Wasseling.

[63] Dumped Chemical Weapons in the Sea – Options, p. 5.

[64] A sternutator is a substance that causes irritation of the nasal and respiratory passages and causes sneezing, coughing, tearing (watering of the eyes), and possibly vomiting.

[65] A vesicant is an agent that produces blisters.

[66] A lachrymator is an irritant that causes tearing.

[67] Stockholm International Peace Research Institute, The Challenge of Old Chemical Munitions and Toxic Armament Wastes, pp. 86-89.

[68] Variations of Z-OA include: 1) 49% Sulfur Mustard, 23% Phenyldichloroarsine (Pfifficus), 19% Diphenylchloroarsine (Clark I), 6% Tornesit (Chlorinated Rubber), 3% S-Wax and 2) Sample from Baltic Sea: 40% Sulfur Mustard, 36.4% Phenyldichloroarsine (Pfifficus), 9% S-Wax, 4.5% Diphenylchloroarsine (Clark I), 3.6% Oxygen Mustard, 2.7% Triphenylarsine, 1.9% Sesqui Mustard, 1.8% Arsenic Trichloride, and 0.1% Picric Acid.

[69] Stockholm International Peace Research Institute, The Challenge of Old Chemical Munitions and Toxic Armament Wastes, p. 96.

[70] Report to the 15th Meeting of the Helsinki Commission from the Ad-Hoc Working Group on Dumped Chemical Munition (HELCOM CHEMU), January 1994, p.11.

[71] Dumped Chemical Weapons in the Sea – Options, 1999, p. 6.

[72] U.K. Ministry of Defence, Report on Sea Dumping of Chemical Weapons by the United Kingdom in the Skagerrak Waters Post World War Two, 1993, p.5

[73] Public Record Office, Foreign Office (FO) File 1005/1959, "15th Progress Report of the Control Commission for Germany (British Element) Covering the Period 1st - 30th September, 1945", Appendix B, para. 2a.

[74] U.K. Ministry of Defence, Report on Sea Dumping of Chemical Weapons by the United Kingdom in the Skagerrak Waters Post World War Two, p.7

[75] National Report of the Russian Federation, "Complex Analysis of the Hazard Related to the Captured German Chemical Weapon Dumped in the Baltic Sea", October 1993, p. 4.

[76] Public Record Office, War Office (WO) File 32, "SECRET Letter from HQ, BAOR to The War Office, Subj: CW Ammunition Held by the French," February 27, 1946.

[77] Stockholm International Peace Research Institute (SIPRI), The Challenge of Old Chemical Munitions and Toxic Armament Wastes, p. 267.

[78] Report to the 15th Meeting of the Helsinki Commission from the Ad-Hoc Working Group on Dumped Chemical Munition (HELCOM CHEMU), January 1994, p.16.

[79] German Federal Maritime and Hydrographic Agency, Chemical Munitions in the Southern and Western Baltic Sea, May 1993, p. 12.

[80] U.S. Army, Final Report, Captured Toxic Ammunition Program, HQ, Continental Base Section, European Command, Chemical Section, May 22, 1947, p. 1.

[81] U.S. Army, Headquarters European Command, Office of the Chief of Chemical Corps, The History of Captured Enemy

Toxic Munitions in the American Zone, European Theater, May 1945 to June, 1947, Section VI "Operation Davey Jones Locker, Chemical Corps, June 1946".

[82] U.S. Army, Headquarters European Command, Addendum to the History of Captured Enemy Toxic Munitions in the American Zone, European Command, June 1947 to August 1948, January 1949, p. 2.

[83] Ibid.

[84] Ibid.

[85] U.S. Army, Headquarters European Command, Addendum to the History of Captured Enemy Toxic Munitions in the American Zone, European Command, June 1947 to August 1948, January 1949, Description of Picture No. 10.

[86] U.S. Army, Headquarters European Command, Addendum to the History of Captured Enemy Toxic Munitions in the American Zone, European Command, June 1947 to August 1948, January 1949, p. 2.

[87] Ibid.

[88] U.S. Army, Headquarters European Command, Addendum to the History of Captured Enemy Toxic Munitions in the American Zone, European Command, June 1947 to August 1948, January 1949, Description of Picture No. 5.

[89] Ibid.

[90] Ibid, Description of Picture No. 7.

[91] U.S. Army, Chemical Weapons Movement History Compilation, Moves – 1946, p. 4 and ACDA Report, p. 8.

[92] U.S. Army, Meeting Notes – Summary of Some Chemical Munitions Sea Dumps by the United States, Mr. William R. Brankowitz, 30 January 1989.

[93] U.S. Army Center of Military History, The Chemical Warfare Service: Chemicals in Combat, pp. 312-313.

[94] Ibid, p. 122.

[95] U.S. Army, Compilation of U.S. Army Toxic Chemical Warfare Agent Production, Storage, Test, Disposal and Decontamination Operations, 1940-1970, p. 148.

[96] U.S. Army, Inventory of Toxic Chemical and Biological (Toxin) Agents, Munitions, Dumps, and Historical Summary, U.S. Army Edgewood Arsenal, 25 August 1970, p. 38.

[97] U.K. Ministry of Defence, "Report on Sea Dumping of Chemical Weapons by the United Kingdom in the Skagerrak Waters Post World War Two", 1993, p. 7.

[98] Ibid, p. 4.

[99] Ibid.

[100] The seaport of Stavanger is on Stavanger Fjord in SW Norway, 190 miles WSW of Oslo. It had been seized by the Germans on April 9, 1940.

[101] The Kattegat is a broad arm of the North Sea, 40 to 70 miles wide, between Sweden on the east and Jutland, Denmark on the west, connecting with the North Sea through the Skagerrak and with the Baltic Sea through Öresund, the Great Belt, and the Little Belt.

[102] U.K. Ministry of Defence, "Report on Sea Dumping of Chemical Weapons by the United Kingdom in the Skagerrak Waters Post World War Two", 1993, p. 5.

[103] Appendix A, Naval Division Report, CCG(BE) Progress Report #13, 31 July 1945.

[104] Redegjørelse vedrørende de miljø-, sundheds- og sikkerhetsmessige forhold I forbindelse med dumpet giftgasammunition I farvandene omkring Danmark. (An account of the environment, health, and security in connection with the dumped wargas ammunition in waters surrounding Denmark), Miljøstyrelsen, København, 1985.

[105] U.K. Ministry of Defence, "Report on Sea Dumping of Chemical Weapons by the United Kingdom in the Skagerrak Waters Post World War Two", 1993, pp. 5-6.

[106] Ibid, p. 6.

[107] Public Record Office, Foreign Office (FO) File 1005/1958, "14th Progress Report of the Control Commission for Germany (British Element) Covering the Period 1st-31st August, 1945."

[108] Hamburger Abendblott, "Gasing Geleilzuge Voll Gasgranaten Versenket", August 19, 1970.

[109] Public Record Office, War Office (WO) File 32/21200, Letter, classified SECRET, from HQ, BAOR to The War Office, Subject: Disposal of Enemy CW Stocks, January 24, 1946.

[110] U.K. Ministry of Defence, R. Bowles, "Standard Reply to Enquiries Re Sea Dumping of Munitions".

[111] Ibid.

[112] Ibid.

[113] U.K. House of Commons Harsard Debates, 30 March 1995, Column 753.

[114] Bulletin of the Atomic Scientists, September/October 1997, Volume 53, Number 5, "A Sea of Trouble" by Ron Chepesiuk.

[115] U.K. Ministry of Defence, R. Bowles, "Standard Reply to Enquiries Re Sea Dumping of Munitions".

[116] Report to the 15th Meeting of the Helsinki Commission from the Ad-Hoc Working Group on Dumped Chemical Munition (HELCOM CHEMU), January 1994, p.17.

[117] Dumped Chemical Weapons in the Sea - Options, 1999, p. 6.

[118] German Federal Maritime and Hydrographic Agency, Chemical Munitions in the Southern and Western Baltic Sea, May 1993, p. 55.

[119] Danish National Environmental Agency, "Detailed Report on the Environmental, Health, and Security Issues Associated with the Dumping of Poison Gas Ammunition in Danish Waters", 7 May 1985, Report from Working Group No. 1, pp. 4-5.

[120] National Report of the Russian Federation, "Complex Analysis of the Hazard Related to the Captured German Chemical Weapon Dumped in the Baltic Sea", October 1993, pp. 6-8.

[121] Mikhail Sergeyevich Gorbachev served as General Secretary of the Communist Party of the Soviet Union from 1985 until 1991, and as the last head of state of the Soviet Union, from 1988 until its dissolution in 1991.

[122] St. Petersburg Chas Pik, "Chas Pik Investigation: Secrets of Chemical Warfare", 17 March 1993. The article included an interview with Dr. of Chemical Sciences Lev Aleksandrovich Fedorov.

[123] German Federal Ministry of Transport, Federal Maritime and Hydrographic Agency, "Chemical Munitions in the Southern and Western Baltic Sea", May 1993, p. 12.

[124] Ibid, p. 13.

[125] Danish National Environmental Agency, "Detailed Report on the Environmental, Health, and Security Issues Associated with the Dumping of Poison Gas Ammunition in Danish Waters", 7 May 1985, Report from Working Group No. 1, p. 12.

[126] German Federal Ministry of Transport, Federal Maritime and Hydrographic Agency, "Chemical Munitions in the Southern and Western Baltic Sea", May 1993, p. 12.

[127] Danish National Environmental Agency, "Detailed Report on the Environmental, Health, and Security Issues Associated with the Dumping of Poison Gas Ammunition in Danish Waters", 7 May 1985, p. 9 and the Report from Working Group No. 1, p. 12.

[128] As stated in paragraph 3.2.4, in 1959-60, German authorities brought two ships that were scuttled in the Baltic's Little Belt to the surface, removed 69,000 tabun nerve gas shells, encapsulated them in concrete and then re-dumped them in the Bay of Biscay in March 1960.

[129] Telegram dated May 8, 1969, submitted for the record by Representative Richard D. McCarthy during the 13 May 1969 Hearing on the International Implications of Dumping Poisonous Gas and Waste into Oceans before the Subcommittee on International Organizations and Movements, Committee on Foreign Affairs, U.S. House of Representatives.

[130] National Report of the Russian Federation, "Complex Analysis of the Hazard Related to the Captured German Chemical Weapon (sic) Dumped in the Baltic Sea", October 1993, p. 8.

[131] Ibid, pp. 6-8.

[132] Ibid.

[133] Ad Hoc Working Group on Dumped Chemical Munition (HELCOM CHEMU), "Report to the 15th Meeting of the

Helsinki Commission", January 1994, p. 11.

[134] Stockholm International Peace Research Institute, "The Challenge of Old Chemical Munitions and Toxic Armament Wastes", 1997, p. 269.

[135] Imperial College Consultants, "Munitions Dumped at Sea: A Literature Review", June 2005, pp. 3-4.

[136] Shirshov Institute of Oceanology, Atlantic Branch, Kaliningrad, Russia, Dr. Vadim, Paka, "Deployment, Toxicity, and Influence on the Environment and other Issues Connected with Sea-Dumped Chemical Weapons", p. 1.

[137] OSPAR Commission, Quality Status Report (QSR) 2010, "Assessment of the Impact of Dumped Conventional and Chemical Munitions: Measures taken by OSPAR." See also "Status and Trend of Marine Chemical Pollution", OSPAR Commission, London 2009, Publication 365/2009.

[138] German Federal Ministry of Transport, Federal Maritime and Hydrographic Agency, Chemical Munitions in the Southern and Western Baltic Sea, p. 26.

[139] Phillip Knightley, "Dumps of Death," *London Times*, April 5, 1992, pp. 26-30.

[140] Statement by Mr. V. Sherbakov, Vice Mayor of St. Petersburg, Russia.

[141] Fredrik Laurin, "Scandinavia's Underwater Time Bomb," *The Bulletin of the Atomic Scientists*, March 1991, pp. 11-15.

[142] Yevgeniy Solomenko, "German Chemical Weapons Dumped in Baltic after World War II," *Moscow Izvestiya*, June 4, 1992, p. 7.

[143] Leonid Pleshakov, "Further Revelations on World War II Chemical Dumping in Baltic," *Moscow Ogonek*, February 8,

1992, pp. 8-9.

[144] German nerve gases were originally referred to as "G Agents" after the marking on the weapons containing tabun. In addition to the color/number coding which indicated the broad characteristics of the filling of German chemical weapons, a code letter was included for more precise identification. "G" indicated a pure tabun filling. "GA" indicated a tabun filling diluted with 20% chlorobenzene. "GA" has since become the standard US code for tabun. See SIPRI, The Problem of Chemical and Biological Warfare, Vol. 1, p. 73.

[145] Public Record Office, Foreign Office (FO) File 1005/1560, "Minutes of the First Meeting Held in the Sixth Floor Conference Room, Norfolk House, on Monday, the 4th December, 1944, at 1500 Hours."

[146] Ibid.

[147] Ibid.

[148] Ibid.

[149] Public Record Office, Foreign Office (FO) File 1005/1561, "Minutes of the Fifth Meeting Held in Room 311, Norfolk House, on Wednesday, 3rd January, 1945, at 1430 Hours", p.3, para. 4.

[150] Public Record Office, Foreign Office (FO) File 1005/1561, "Minutes of the Sixth Meeting Held in Room 311, Norfolk House, on Wednesday, 10th January, 1945, at 1500 Hours", p.3, para.12.

[151] Public Record Office, Foreign Office (FO) File 1005/1561, "Minutes of the Seventh Meeting held in Room 311, Norfolk House, on Wednesday, 24th January, 1945, at 1500 Hours", p.3, para.21.

[152] Public Record Office, Foreign Office (FO) File 1005/1561,

---

"Minutes of Meeting of Chairmen of Sub-Committees held in Room 126, Norfolk House, at 1030 hrs, 14th February, 1945", para.1.

[153] U.K. Ministry of Defence, "Report on Sea Dumping of Chemical Weapons by the United Kingdom in the Skagerrak Waters Post World War Two", 1993, p. 1.

[154] National Archives of the United States, formerly CONFIDENTIAL Voyage Report of the S.S. Cape Borda from Belfast, N. Ireland to Philadelphia, Pennsylvania, June 13, 1945.

[155] Public Record Office, Admiralty (ADM) File 1/18116, "Unserviceable Ammunition to be Dumped at Sea," June 25, 1945, Minute Sheet No. 4, Register No. M.03124/45.

[156] Public Record Office, Foreign Office (FO) File 1005/1956, "Twelfth Progress Report of the Control Commission for Germany (British Element) Covering the Period 1st-30th June, 1945".

[157] U.K. Ministry of Defence, "Report on Sea Dumping of Chemical Weapons by the United Kingdom in the Skagerrak Waters Post World War Two", 1993, p. 5.

[158] Public Record Office, Admiralty (ADM) File 1/18116, "Unserviceable Ammunition to be Dumped at Sea," July 11, 1945, Minute Sheet No. 6.

[159] U.K. Ministry of Defence, "Report on Sea Dumping of Chemical Weapons by the United Kingdom in the Skagerrak Waters Post World War Two", 1993, p. 4.

[160] Ibid.

[161] The seaport of Stavanger is on Stavanger Fjord in SW Norway, 190 miles WSW of Oslo. It had been seized by the Germans on April 9, 1940.

[162] The Kattegat is a broad arm of the North Sea, 40 to 70 miles wide, between Sweden on the east and Jutland, Denmark on the west, connecting with the North Sea through the Skagerrak and with the Baltic Sea through Öresund, the Great Belt, and the Little Belt.

[163] U.K. Ministry of Defence, "Report on Sea Dumping of Chemical Weapons by the United Kingdom in the Skagerrak Waters Post World War Two", 1993, p. 5.

[164] Public Record Office, Foreign Office (FO) File 1005/1957, "13th Progress Report of the Control Commission for Germany (British Element) Covering the Period 1st-31st July, 1945".

[165] Redegjørelse vedrørende de miljø-, sundheds- og sikkerhetsmessige forhold I forbindelse med dumpet giftgasammunition I farvandene omkring Danmark. (An account of the environment, health, and security in connection with the dumped wargas ammunition in waters surrounding Denmark), Miljøstyrelsen, København, 1985.

[166] Bremen was a submarine and naval base during WWII, was largely destroyed by Allied bombing in 1943-45, and was taken by the British on April 27, 1945.

[167] The Skagerrak is a broad arm of the North Sea, about 150 miles long and 80 miles wide, extending between Norway on the north and Denmark on the south and connecting with the Kattegat on the east.

[168] Paragraph IIA3(i)(b), Potsdam Accord and MoD, Documents on British Policy Overseas, Series I, Volume I, No. 603.

[169] Norwegian Defence Research Establishment (NDRE), Forsvarets Forskningsinstitutt (FFI), An Investigation into the

Sunken Chemical Ships Outside the Norwegian Coast, Professor Frode Fonnum, 1993, p. 1.

170 U.K. Ministry of Defence, "Report on Sea Dumping of Chemical Weapons by the United Kingdom in the Skagerrak Waters Post World War Two", 1993, p. 4.

171 Public Record Office, Foreign Office (FO) File 1005/1561, "Minutes of the 18th Meeting held in the Chief of Staff's Conference Room, Koenigshof, Bad Oeynhausen on Thursday, 16th August, 1945, at 1500 Hours", p.26, para. 98(1).

172 U.K. Ministry of Defence, "Report on Sea Dumping of Chemical Weapons by the United Kingdom in the Skagerrak Waters Post World War Two", 1993, p. 3.

173 Teleprint dated August 20, 1945 from Searail 7 to the Chairman, Continental Ammunition Dumping Committee.

174 Public Record Office, War Office (WO) File 32/21200, formerly CONFIDENTIAL Letter from HQ, 21st Army Group to The Under Secretary of State and The War Office, Subject: Disposal of Chemical Ammunition, August 23, 1945.

175 Public Record Office, War Office (WO) File 32/21200, "Minutes of 4th Meeting of Continental Ammunition Dumping Committee held at HQ BAOR 29th August 1945."

176 Public Record Office, Foreign Office (FO) File 1005/1958, "14th Progress Report of the Control Commission for Germany (British Element) Covering the Period 1st-31st August, 1945."

177 Public Record Office, Foreign Office (FO) File 1005/1561, "Minutes of the 19th Meeting held in the Chief of Staff's Conference Room, Koenigshof, Bad Oeynhausen on Monday, 3rd September, 1945, at 1500 Hours", para. 103 (1).

[178] U.S. Memoranda Concerning Captured WWII CW Material, Appendix I to Special Study on the Sea Disposal of Chemical Munitions, United States Arms Control and Disarmament Agency, January 19, 1993.

[179] Public Record Office, Foreign Office (FO) File 1005/1561, "Minutes of the 20th Meeting held in the Chief of Staff's Conference Room, Koenigshof, Bad Oeynhausen on Monday, 17th September, 1945, at 1500 Hours", p. 1, para. 109 (1).

[180] Public Record Office, War Office (WO) File 32/21200, "Minutes of 5th Meeting of Continental Ammunition Dumping Committee held at HQ, BAOR 17th September 1945."

[181] Public Record Office, Foreign Office (FO) File 1005/1959, "15th Progress Report of the Control Commission for Germany (British Element) Covering the Period 1st - 30th September, 1945", Appendix A.

[182] Public Record Office, Foreign Office (FO) File 1005/1959, "15th Progress Report of the Control Commission for Germany (British Element) Covering the Period 1st - 30th September, 1945", Appendix B, para. 2a.

[183] Public Record Office, Foreign Office (FO) File 1005/1959, "15th Progress Report of the Control Commission for Germany (British Element) Covering the Period 1st - 30th September, 1945", Appendix D, Annex XX.

[184] Public Record Office, Foreign Office (FO) File 1005/1561, "Minutes of the 21st Meeting held in the Chief of Staff's Conference Room, Koenigshof, Bad Oeynhausen on Monday, 4th October, 1945, at 1500 Hours", p. 2, para. 117 (1).

[185] Public Record Office, War Office (WO) File 32/21200, "Minutes of Sixth Meeting of Continental Ammunition Dumping Committee held at HQ BAOR, 5 October 1945, para 3, p. 2.

[186] Public Record Office, War Office (WO) File 32/21200, formerly TOP SECRET telegram (JSM 73) from the Joint Staff Mission, Washington, DC to the Cabinet Offices, October 5, 1945.

[187] Public Record Office, War Office (WO) File 32/21200, formerly RESTRICTED Letter from HQ, BAOR to The War Office, Subject: Disposal of Enemy CW Ammunition, October 11, 1945.

[188] Public Record Office, War Office (WO) File 32/21200, formerly TOP SECRET letter (BM 5171 SWV2) from The War Office to HQ, BAOR, October 13, 1945.

[189] Public Record Office, Foreign Office (FO) File 1005/1960, "16th Progress Report of the Control Commission for Germany (British Element) Covering the Period 1st - 31st October, 1945", Appendix A, p. 3, para. 7.

[190] Ibid, Appendix B, p. 1, para. 2a.

[191] Ibid, para. 2b.

[192] Ibid, Appendix D, Annexes XI and XII.

[193] Hamburger Abendblott, "Gasing Geleilzuge Voll Gasgranaten Versenket", August 19, 1970.

[194] Public Record Office, War Office (WO) File 32/21200, "Minutes of the 7th Meeting of Continental Ammunition Dumping Committee", Report No. BAOR/18581/Q(M)3, December 7, 1945.

[195] Letter, Subject: Disposition of Ammunition, Brigadier General Henry J. Wolfe, U.S. Army, Acting Director, Plans and Operations, ASF, December 27, 1945.

[196] U.S. Memoranda Concerning Captured WWII CW Material, Appendix I to Special Study on the Sea Disposal of Chemical Munitions, United States Arms Control and Disarmament

Agency, January 19, 1993.

197 U.S. Memoranda Concerning Captured WWII CW Material, Appendix I to Special Study on the Sea Disposal of Chemical Munitions, United States Arms Control and Disarmament Agency, January 19, 1993.

198 Ibid.

199 Public Record Office, War Office (WO) File 32/21200, Letter, classified SECRET, from HQ, BAOR to The War Office, Subject: Disposal of Enemy CW Stocks, January 24, 1946.

200 Public Record Office, Foreign Office (FO) File 1005/1562, "Minutes of the 27th Meeting held in the War Establishments Committee Room, 2A Porta Strasse, Bad Oeynhausen on Thursday, 24th January, 1946, at 1500 Hours", p. 4, para. 178.

201 U.S. Army, Captured Enemy Material, Office of the Chief Historian, European Command, 1947, p. 27.

202 Public Record Office, War Office (WO) File 32, "SECRET Letter from HQ, BAOR to The War Office, Subj: CW Ammunition Held by the French," February 27, 1946.

203 Public Record Office, Foreign Office (FO) File 1005/1562, "Minutes of the 28th Meeting held in the Chief of Staff's Conference Room, Koenigshof Hotel, Bad Oeynhausen on Thursday, 21st March, 1946, at 1500 Hours", p. 2, para. 182.

204 U.S. Army, Captured Enemy Ammunition, Office of the Chief Historian, European Command, 1947, p. 29.

205 Ibid, p. 28.

206 Ibid.

207 U.S. Army, Final Report, Captured Toxic Ammunition Program, HQ, Continental Base Section, European Command, Chemical Section, May 22, 1947, p. 1.

[208] Norwegian Defence Research Establishment (NDRE), Forsvarets Forskningsinstitutt (FFI), An Investigation into the Sunken Chemical Ships Outside the Norwegian Coast, Professor Frode Fonnum, 1993, p. 3.

[209] U.S. Army (1947), The History of Captured Enemy Toxic Munitions in the American Zone, European Theater, May 1945 to June 1947, Section I.

[210] Public Record Office, Foreign Office (FO) File 1005/488.

[211] The seaport of Emden in NW Germany was an important naval base with oil tanks and refineries during WWII. It was frequently heavily bombed by the Allies in 1943-45 and was taken by Canadian forces in April 1945.

[212] Public Record Office, Foreign Office (FO) File 1005/488, Minutes of the Second Meeting, para. 13.

[213] Public Record Office, Foreign Office (FO) File 1005/488, Minutes of the Third Meeting, para. 20.

[214] Public Record Office, Foreign Office (FO) File 1005/488, Minutes of the Fifth Meeting, para.31.

[215] Public Record Office, Foreign Office (FO) File 1005/496, Report of the Prime Minister's Commission on Demilitarization in the British Zone of Germany, p. 3, para. B4.

[216] Public Record Office, Foreign Office (FO) File 1005/496, Report of the Prime Minister's Commission on Demilitarization in the British Zone of Germany, p. 3, para. B6.

[217] Ibid, para. 19.

[218] Public Record Office, Foreign Office (FO) File 1005/496, Appendix D to Report of the Prime Minister's Commission on Demilitarization in the British Zone of Germany, "Progress

Report on Disposal of Certain Articles of German War Material in the British Zone of Occupation in Germany during the Period from May 1945 to 1 May 1946", p. 2.

[219] U.K. Ministry of Defence, "Report on Sea Dumping of Chemical Weapons by the United Kingdom in the Skagerrak Waters Post World War Two", 1993, p. 4.

[220] Public Record Office, Foreign Office (FO) File 1005/489, Minutes of the Seventh Meeting, para. 19.

[221] U.K. Ministry of Defence, "Report on Sea Dumping of Chemical Weapons by the United Kingdom in the Skagerrak Waters Post World War Two", 1993, p. 7.

[222] Ibid.

[223] Ibid.

[224] Helsinki Commission, Ad Hoc Working Group on Dumped Chemical Munition, "Report on Chemical Munitions Dumped in the Baltic Sea," January 1994, p. 11 and the German Federal Ministry of Transport, Federal Maritime and Hydrographic Agency, "Chemical Munitions in the Southern and Western Baltic Sea," May 1993, p. 7. The original German source document from the Federal Archives in Koblenz referenced by the German Ministry of Transport is "Ferigung, Lagerung and Beseitigung Chemischer Kampfstoffe unter Besonderer. Berücksichtigung des Territoriums der Bundesrepublik Deutschland. Ein Bericht des Bundesarchivs; Koblenz und Freiburg," 1979.

[225] Public Record Office, Foreign Office (FO) File 1005/490, Minutes of the 18th Meeting, p. 11.

www.ingramcontent.com/pod-product-compliance
Lightning Source LLC
Chambersburg PA
CBHW051437170526
45166CB00001B/23